It's another Quality Book from CGP

This book is for anyone doing GCSE AQA A Geography.

Whatever subject you're doing it's the same
old story — there are lots of facts and you've just got
to learn them. KS4 Geography is no different.

Happily this CGP book gives you all that important
information as clearly and concisely as possible.

It's also got some daft bits in to try and make the whole
experience at least vaguely entertaining for you.

What CGP is all about

Our sole aim here at CGP is to produce the highest quality
books — carefully written, immaculately presented and
dangerously close to being funny.

Then we work our socks off to get them out to you
— at the cheapest possible prices.

Contents

Published by CGP

Editors:
Ellen Bowness, Joe Brazier, Taissa Csaky, Chris Dennett, Murray Hamilton, Paul Jordin,
Kate Redmond, Hayley Thompson, Jane Towle, Karen Wells.

Contributors:
Caroline Batten, Rosalind Browning, Polly Cotterill, Paddy Gannon, Neil Hastings, Helen Marr,
Barbara Melbourne, Sophie Watkins, Dennis Watts.

Proofreading:
Simon Little, Eileen Worthington.

ISBN: 978 1 84762 378 2

With thanks to Laura Phillips for copyright research.

With thanks to iStockphoto.com for permission to reproduce the photographs used on pages 7, 17, 24, 32, 58 and 89.

With thanks to Science Photo Library for permission to reproduce the photographs used on pages 8, 18 and 73.

Images of Sri Lankan coastline on page 11 © UPPA/Photoshot.

Map of UK geology on page 13 reproduced by permission of the British Geological Survey.
© NERC. All rights reserved. IPR/115-58CT.

Graphs of rainfall and sunshine hours on page 22 adapted from Crown Copyright data supplied by the Met Office.

Graph of the last 1000 years of climate change on page 25 reproduced with kind permission from Climate Change 2001:
The Scientific Basis, Contribution of Working Group I to the Third Assessment Report of the Intergovernmental Panel on
Climate Change, SPM Figure 1. Cambridge University Press.

Graph of the last 150 years of climate change on pages 25 and 56 adapted from Crown Copyright data supplied by the
Met Office.

Mapping data on pages 46, 59, 68, 144 and 145 reproduced by permission of Ordnance Survey® on behalf
of HMSO © Crown copyright (2009). All rights reserved. Ordnance Survey® Licence No. 100034841.

Data used to compile the UK population density maps on pages 51 and 139 from Office for National Statistics: General
Register Office for Scotland, Northern Ireland Statistics & Research Agency. © Crown copyright reproduced under the
terms of the Click-Use Licence.

Data used to compile the UK average rainfall map on page 51 from the Manchester Metropolitan University.

Image of Rhône Glacier in 2008 on page 56 © Juerg Alean, Eglisau, Switzerland, http://www.glaciers-online.net.

Data used to produce Rhône Glacier graph on page 56 © VAW / ETH Zurich,
http://www.glaciology.ethz.ch/swiss-glaciers/

Data used to construct the UK population pyramid on page 81 © Crown copyright reproduced under
the terms of the Click-Use Licence.

World Population Graph on page 85 reproduced with kind permission from Jean-Paul Rodrigue
(underlying data from the United Nations).

With thanks to Mr Steve Ellingham for the image of trekking in the Himalayas on page 127.

Data used to construct the flow map of immigration on page 143 - Source International Passenger Survey,
Office for National Statistics © Crown copyright reproduced under the terms of the Click-Use Licence.

Every effort has been made to locate copyright holders and obtain permission to reproduce sources.
For those sources where it has been difficult to trace the copyright holder of the work, we would be grateful
for information. If any copyright holder would like us to make an amendment to the acknowledgements,
please notify us and we will gladly update the book at the next reprint. Thank you.

Groovy website: www.cgpbooks.co.uk
Printed by Elanders Ltd, Newcastle upon Tyne.
Jolly bits of clipart from CorelDRAW®

Based on the classic CGP style created by Richard Parsons.

Structure of the Course

'Know thy enemy', 'forewarned is forearmed'... There are many boring quotes that just mean <u>being prepared is a good thing</u>. <u>Don't</u> stumble <u>blindly</u> into a GCSE course — find out what you're facing.

You'll have to do Two Exams at Either Higher or Foundation Level

GCSE Geography's divided into <u>three units</u> — <u>Unit 1: Physical Geography</u>, <u>Unit 2: Human Geography</u> and <u>Unit 3: Local Fieldwork Investigation</u>. You'll have to do <u>two exams</u> — one for <u>Unit 1</u> and one for <u>Unit 2</u>. You'll either be entered for the <u>Higher</u> or the <u>Foundation level</u> exams — they cover the <u>same topics</u> (<u>all</u> covered in this book), but the <u>questions</u> are <u>slightly different</u>.

Unit 1: Physical Geography

Unit 1's divided into <u>two sections</u> (A and B) and <u>seven topics</u>:

> <u>Section A</u>
> * The Restless Earth
> * Rocks, Resources and Scenery
> * Challenge of Weather and Climate
> * The Living World
>
> <u>Section B</u>
> * Water on the Land
> * Ice on the Land
> * The Coastal Zone

Here's how the exam's <u>structured</u>:

🕐 1 hour 30 minutes	75 marks in total	37.5% of your final mark

There are <u>seven</u> questions in total — <u>one on each</u> topic (see above). You need to answer <u>three out of the seven</u> questions — <u>one question</u> from <u>Section A</u>, <u>one question</u> from <u>Section B</u>, then a <u>third question</u> from <u>either</u> section.

Unit 2: Human Geography

Unit 2's divided into <u>two sections</u> (A and B) and <u>six topics</u>:

> <u>Section A</u>
> * Population Change
> * Changing Urban Environments
> * Changing Rural Environments
>
> <u>Section B</u>
> * The Development Gap
> * Globalisation
> * Tourism

Here's how the exam's <u>structured</u>:

🕐 1 hour 30 minutes	75 marks in total	37.5% of your final mark

There are <u>six</u> questions in total — <u>one on each</u> topic (see above). You need to answer <u>three out of the six</u> questions — <u>one question</u> from <u>Section A</u>, <u>one question</u> from <u>Section B</u>, then a <u>third question</u> from <u>either</u> section.

'One question' means a big multi-part beastie worth 25 marks in total.

Unit 3: Local Fieldwork Investigation

The <u>local fieldwork investigation</u> involves some fieldwork (<u>outdoor fun</u>, often in wellies) and a <u>written report</u> — it used to be called <u>coursework</u>. It's done under <u>controlled conditions</u> (a bit like exam conditions).

The fieldwork bit: Can be done on your <u>own</u>, in <u>groups</u> or as a <u>class</u>. It involves collecting <u>primary data</u> — data you collect <u>yourself</u>, e.g. measurements of erosion, questionnaire responses, etc.

The report part: You <u>write up</u> your methods and present your data. Then you <u>describe</u>, <u>analyse</u> and make <u>conclusions</u> about your data.

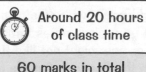

🕐 Around 20 hours of class time
60 marks in total 25% of your final mark
Suggested word limit: 2000

May the course be with you...

It's worthwhile <u>knowing</u> all of this stuff so nothing comes as a <u>shock</u> to you. It also stops you from being the person who tried to answer every single question in the <u>exam</u> — there's a fine line between bravery and self-sabotage...

How to Use this Book

That last page was a bit scary, talking about <u>exams</u> when you haven't even started <u>revising</u> yet. But don't worry — this book's here to help when you do start revising, so you'll ace your exams.

This Book's in the Same Order as the Exams

This book is arranged to follow the <u>structure of the course</u> (see previous page). For example:

> <u>Unit 1A — The Restless Earth</u> — covers all the material you need for Unit 1: Physical Geography, Section A — The Restless Earth.
>
> <u>Unit 2B — Globalisation</u> — covers all the material you need for Unit 2: Human Geography, Section B — Globalisation.

Muuum, I can't sleep...

You Might Not Study All the Topics

1) It sounds odd, but you <u>don't</u> actually <u>have to learn all of the topics</u> on this course.

2) In each exam there's <u>one question on each topic</u> and you only have to answer <u>three questions</u> — one from Section A, one from Section B and one more from either section.

3) So, it <u>may be</u> that you <u>only learn some of the topics</u> — if you're not sure, ask your teacher <u>which topics</u> you should be learning.

4) <u>Circle the topics</u> you've got to <u>revise</u> on the <u>contents page</u> of this book so you can remember — then you'll know <u>which questions to answer</u> in the exam.

Don't scribble on your book if it's a school copy.

Work Out Your Strengths and Weaknesses

1) Make sure you know your <u>three strongest</u> topics in <u>both</u> Unit 1 (physical geography) and Unit 2 (human geography), then look at those questions <u>first</u> in each exam.

2) For example, if you got <u>ALL</u> the test questions right for the Unit 1 topics The Living World, Water on the Land and Ice on the Land, look at those questions <u>first</u>.

3) <u>But remember</u>, you need to answer <u>at least one</u> question from each section, so <u>make sure</u> you have <u>at least one</u> strong topic in <u>Section A</u> and one strong topic in <u>Section B</u> for <u>both units</u>.

4) Make sure you know your <u>weakest</u> topics too — this means you can spend a bit <u>more time</u> revising those topics, so you can <u>answer</u> those questions in the exam if you really have to.

Add to this Book and Practise, Practise, Practise

This book covers <u>all you need to know</u>, but for <u>top marks</u>, there are a couple of <u>other things</u> you can do:

Make your own notes

<u>Add your own notes</u> or put pages of your class notes into this book, e.g. you might not like the <u>case study</u> on the floods in Carlisle and South Asia — so <u>stick</u> your own case study in that place instead (just make sure it <u>covers the same points</u>).

Practise exam questions

1) This book contains <u>everything</u> you need to <u>revise</u>, but you need to do <u>practice exam questions too</u>.

2) There's a handy <u>Exam Skills</u> section that should help with all the different <u>skills</u> you might need in an exam, but you still need to actually <u>practise</u> them.

3) Try having a go at some <u>past exam papers</u> (ask your teacher where to find these).

Practice exams — as if life isn't bad enough already...

Well, now you know <u>what's coming your way</u> and <u>how to use this book</u>, the time has come when you've got to <u>sit down</u> and do the <u>revision</u>. Don't worry though, I'll be here every step of the way to keep you <u>entertained</u>.

Tectonic Plates

The Earth's <u>surface</u> is made of huge floating <u>plates</u> that are constantly moving... Rock on.

The Earth's Surface is Separated into Tectonic Plates

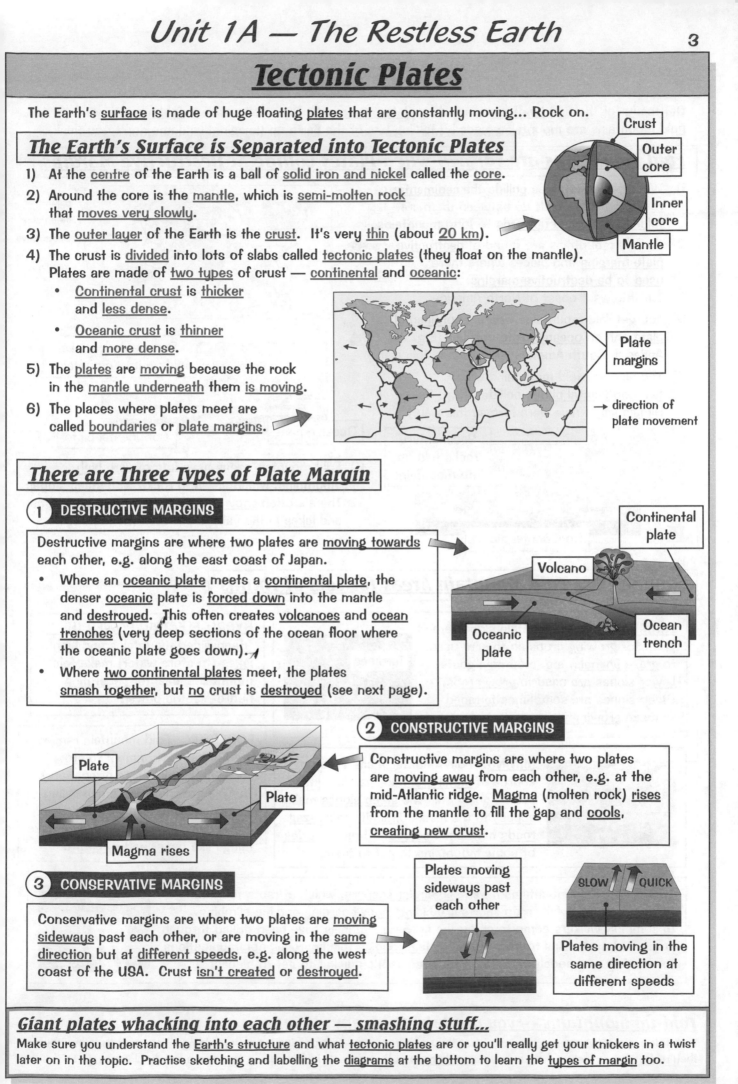

1) At the <u>centre</u> of the Earth is a ball of <u>solid iron and nickel</u> called the <u>core</u>.

2) Around the core is the <u>mantle</u>, which is <u>semi-molten rock</u> that <u>moves very slowly</u>.

3) The <u>outer layer</u> of the Earth is the <u>crust</u>. It's very <u>thin</u> (about <u>20 km</u>).

4) The crust is <u>divided</u> into lots of slabs called <u>tectonic plates</u> (they float on the mantle). Plates are made of <u>two types</u> of crust — <u>continental</u> and <u>oceanic</u>:

- <u>Continental crust</u> is <u>thicker</u> and <u>less dense</u>.
- <u>Oceanic crust</u> is <u>thinner</u> and <u>more dense</u>.

5) The <u>plates</u> are <u>moving</u> because the rock in the <u>mantle underneath</u> them <u>is moving</u>.

6) The places where plates meet are called <u>boundaries</u> or <u>plate margins</u>.

Crust
Outer core
Inner core
Mantle

Plate margins

→ direction of plate movement

There are Three Types of Plate Margin

① DESTRUCTIVE MARGINS

Destructive margins are where two plates are <u>moving towards</u> each other, e.g. along the east coast of Japan.

- Where an <u>oceanic plate</u> meets a <u>continental plate</u>, the denser <u>oceanic</u> plate is <u>forced down</u> into the mantle and <u>destroyed</u>. This often creates <u>volcanoes</u> and <u>ocean trenches</u> (very deep sections of the ocean floor where the oceanic plate goes down).
- Where <u>two continental plates</u> meet, the plates <u>smash together</u>, but <u>no</u> crust is <u>destroyed</u> (see next page).

Continental plate
Volcano
Ocean trench
Oceanic plate

② CONSTRUCTIVE MARGINS

Constructive margins are where two plates are <u>moving away</u> from each other, e.g. at the mid-Atlantic ridge. <u>Magma</u> (molten rock) <u>rises</u> from the mantle to fill the gap and <u>cools</u>, <u>creating new crust</u>.

Plate
Plate
Magma rises

③ CONSERVATIVE MARGINS

Conservative margins are where two plates are <u>moving sideways</u> past each other, or are moving in the <u>same direction</u> but at <u>different speeds</u>, e.g. along the west coast of the USA. Crust <u>isn't created</u> or <u>destroyed</u>.

Plates moving sideways past each other

SLOW QUICK

Plates moving in the same direction at different speeds

Giant plates whacking into each other — smashing stuff...

Make sure you understand the <u>Earth's structure</u> and what <u>tectonic plates</u> are or you'll really get your knickers in a twist later on in the topic. Practise sketching and labelling the <u>diagrams</u> at the bottom to learn the <u>types of margin</u> too.

Fold Mountains

Get ready for the first <u>landform</u> created by <u>plate movement</u>. Drum roll please... <u>fold mountains</u> — ta dah. Fold mountains are mountains made by folding bits of the Earth up (imaginative name don't you think).

Fold Mountains are Formed when Plates Collide at Destructive Margins

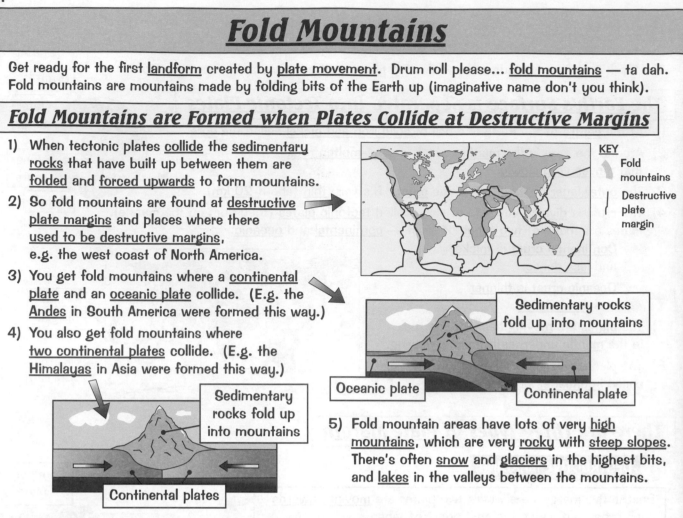

1) When tectonic plates <u>collide</u> the <u>sedimentary rocks</u> that have built up between them are <u>folded</u> and <u>forced upwards</u> to form mountains.

2) So fold mountains are found at <u>destructive plate margins</u> and places where there <u>used to be destructive margins</u>, e.g. the west coast of North America.

3) You get fold mountains where a <u>continental plate</u> and an <u>oceanic plate</u> collide. (E.g. the <u>Andes</u> in South America were formed this way.)

4) You also get fold mountains where <u>two continental plates</u> collide. (E.g. the <u>Himalayas</u> in Asia were formed this way.)

KEY
Fold mountains
Destructive plate margin

Sedimentary rocks fold up into mountains

Oceanic plate   Continental plate

Sedimentary rocks fold up into mountains

Continental plates

5) Fold mountain areas have lots of very <u>high mountains</u>, which are very <u>rocky</u> with <u>steep slopes</u>. There's often <u>snow</u> and <u>glaciers</u> in the highest bits, and <u>lakes</u> in the valleys between the mountains.

Humans Use Fold Mountain Areas for Lots of Things

<u>FARMING</u>: <u>Higher</u> mountain slopes aren't great for growing crops so they're used to <u>graze animals</u>, e.g. mountain goats. <u>Lower</u> slopes are used to <u>grow crops</u>. Steep slopes are sometimes <u>terraced</u> to make <u>growing crops easier</u>.

Terraces

<u>HYDRO-ELECTRIC POWER (HEP)</u>: <u>Steep-sided</u> mountains and <u>high lakes</u> (to store water) make fold mountains ideal for <u>generating hydro-electric power</u>.

<u>MINING</u>: Fold mountains are a major source of <u>metal ores</u>, so there's a lot of mining going on. The <u>steep slopes</u> make <u>access</u> to the mines <u>difficult</u>, so <u>zig-zag roads</u> have been <u>carved</u> out on the <u>sides</u> of some mountains to get to them.

<u>FORESTRY</u>: Fold mountain ranges are a good environment to <u>grow</u> some types of <u>tree</u> (e.g. conifers). They're grown on the steep valley slopes and are used for things like <u>fuel</u>, <u>building materials</u>, and to make things like <u>paper</u> and <u>furniture</u>.

<u>TOURISM</u>: Fold mountains have <u>spectacular scenery</u>, which attracts tourists. In <u>winter</u>, people visit to do sports like <u>skiing</u>, <u>snowboarding</u> and <u>ice climbing</u>. In <u>summer</u>, <u>walkers</u> come to enjoy the scenery. <u>Tunnels</u> have been drilled through some fold mountains to make <u>straight</u>, <u>fast roads</u>. This <u>improves communications</u> for tourists and people who live in the area as it's quicker to get to places.

Fold-up mountains — you can take them anywhere...

Yep, fold mountains are pretty much what they say they are — <u>mountains</u> made by <u>folding</u>. You need to know how they're <u>formed</u> as well as how humans <u>use them</u>, and that doesn't include using them as a handy place to go to the loo.

Fold Mountains — Case Study

Brace yourself for the first <u>case study</u> of many. It's quite a groovy one about the <u>Alps</u> though. Make sure you learn the <u>specific facts</u> for good marks in the exam.

The Alps is a Fold Mountain Range

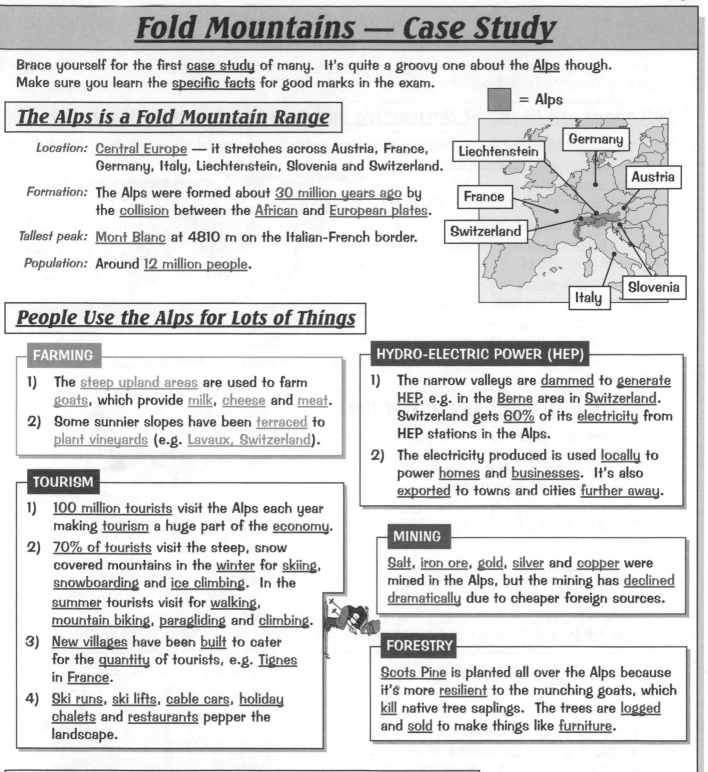

☐ = Alps

Location: <u>Central Europe</u> — it stretches across Austria, France, Germany, Italy, Liechtenstein, Slovenia and Switzerland.

Formation: The Alps were formed about <u>30 million years ago</u> by the <u>collision</u> between the <u>African</u> and <u>European plates</u>.

Tallest peak: <u>Mont Blanc</u> at 4810 m on the Italian-French border.

Population: Around <u>12 million people</u>.

People Use the Alps for Lots of Things

FARMING

1) The <u>steep upland areas</u> are used to farm <u>goats</u>, which provide <u>milk</u>, <u>cheese</u> and <u>meat</u>.
2) Some sunnier slopes have been <u>terraced</u> to <u>plant vineyards</u> (e.g. <u>Lavaux, Switzerland</u>).

TOURISM

1) <u>100 million tourists</u> visit the Alps each year making <u>tourism</u> a huge part of the <u>economy</u>.
2) <u>70% of tourists</u> visit the steep, snow covered mountains in the <u>winter</u> for <u>skiing</u>, <u>snowboarding</u> and <u>ice climbing</u>. In the <u>summer</u> tourists visit for <u>walking</u>, <u>mountain biking</u>, <u>paragliding</u> and <u>climbing</u>.
3) <u>New villages</u> have been <u>built</u> to cater for the <u>quantity</u> of tourists, e.g. <u>Tignes</u> in <u>France</u>.
4) <u>Ski runs</u>, <u>ski lifts</u>, <u>cable cars</u>, <u>holiday chalets</u> and <u>restaurants</u> pepper the landscape.

HYDRO-ELECTRIC POWER (HEP)

1) The narrow valleys are <u>dammed</u> to <u>generate HEP</u>, e.g. in the <u>Berne</u> area in <u>Switzerland</u>. Switzerland gets <u>60%</u> of its <u>electricity</u> from HEP stations in the Alps.
2) The electricity produced is used <u>locally</u> to power <u>homes</u> and <u>businesses</u>. It's also <u>exported</u> to towns and cities <u>further away</u>.

MINING

<u>Salt</u>, <u>iron ore</u>, <u>gold</u>, <u>silver</u> and <u>copper</u> were mined in the Alps, but the mining has <u>declined dramatically</u> due to cheaper foreign sources.

FORESTRY

<u>Scots Pine</u> is planted all over the Alps because it's more <u>resilient</u> to the munching goats, which <u>kill</u> native tree saplings. The trees are <u>logged</u> and <u>sold</u> to make things like <u>furniture</u>.

People Have Adapted to the Conditions in the Alps

1) <u>STEEP RELIEF:</u> <u>Goats</u> are <u>farmed</u> there because they're <u>well adapted</u> to live on <u>steep mountains</u>. <u>Trees</u> and <u>man-made defences</u> are used to <u>protect</u> against <u>avalanches</u> and <u>rock slides</u>.
2) <u>POOR SOILS:</u> <u>Animals</u> are <u>grazed</u> in <u>most high areas</u> as the soil isn't great for growing crops.
3) <u>LIMITED COMMUNICATIONS:</u> <u>Roads</u> have been built over <u>passes</u> (lower points between mountains), e.g. the <u>Brenner Pass</u> between Austria and Italy. It takes a <u>long time</u> to drive over passes and they can be <u>blocked by snow</u>, so <u>tunnels</u> have been cut through the mountains to provide <u>fast transport links</u>. For example, the <u>Lötschberg Base Tunnel</u> has been cut through the Bernese Alps in <u>Switzerland</u>.

Steep relief — the feeling you'll get after you've done your exams...

As with all case studies learn the <u>facts and figures</u> — the examiners go all giddy and throw marks at you when they read them. Learn <u>how people use the Alps</u> and how people have <u>adapted</u> to the <u>conditions</u> that exist there.

Volcanoes

Volcanoes usually look like mountains... until they <u>explode</u> and throw <u>molten rock</u> everywhere. Honestly, they've got such a temper, though I'd probably be in a bad mood if I was sitting on a load of <u>magma</u>.

Volcanoes are Found at Destructive and Constructive Plate Margins

1) At <u>destructive plate margins</u> the <u>oceanic plate</u> <u>goes under</u> the <u>continental plate</u> because it's <u>more dense</u>. (This also creates an <u>ocean trench</u>.):

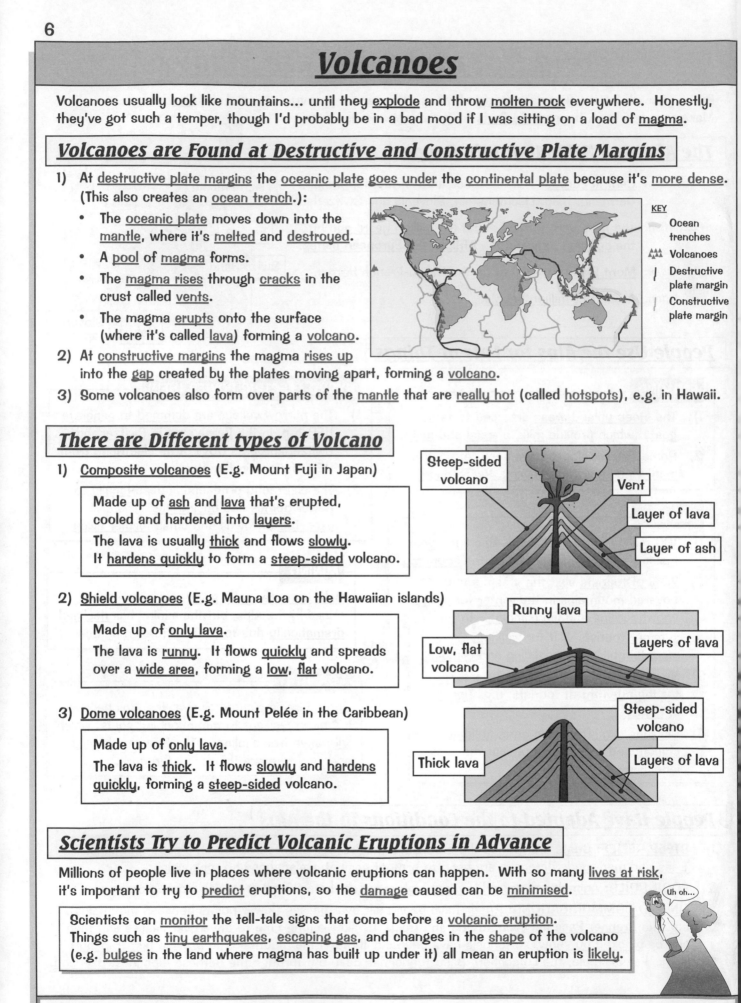

- The <u>oceanic plate</u> moves down into the <u>mantle</u>, where it's <u>melted</u> and <u>destroyed</u>.
- A <u>pool</u> of <u>magma</u> forms.
- The <u>magma rises</u> through <u>cracks</u> in the crust called <u>vents</u>.
- The magma <u>erupts</u> onto the surface (where it's called <u>lava</u>) forming a <u>volcano</u>.

KEY
~ Ocean trenches
▲▲▲ Volcanoes
| Destructive plate margin
| Constructive plate margin

2) At <u>constructive margins</u> the magma <u>rises up</u> into the <u>gap</u> created by the plates moving apart, forming a <u>volcano</u>.

3) Some volcanoes also form over parts of the <u>mantle</u> that are <u>really hot</u> (called <u>hotspots</u>), e.g. in Hawaii.

There are Different types of Volcano

1) <u>Composite volcanoes</u> (E.g. Mount Fuji in Japan)

Made up of <u>ash</u> and <u>lava</u> that's erupted, cooled and hardened into <u>layers</u>.

The lava is usually <u>thick</u> and flows <u>slowly</u>. It <u>hardens quickly</u> to form a <u>steep-sided</u> volcano.

Steep-sided volcano · Vent · Layer of lava · Layer of ash

2) <u>Shield volcanoes</u> (E.g. Mauna Loa on the Hawaiian islands)

Made up of <u>only lava</u>.

The lava is <u>runny</u>. It flows <u>quickly</u> and spreads over a <u>wide area</u>, forming a <u>low</u>, <u>flat</u> volcano.

Runny lava · Layers of lava · Low, flat volcano

3) <u>Dome volcanoes</u> (E.g. Mount Pelée in the Caribbean)

Made up of <u>only lava</u>.

The lava is <u>thick</u>. It flows <u>slowly</u> and <u>hardens quickly</u>, forming a <u>steep-sided</u> volcano.

Steep-sided volcano · Thick lava · Layers of lava

Scientists Try to Predict Volcanic Eruptions in Advance

Millions of people live in places where volcanic eruptions can happen. With so many <u>lives at risk</u>, it's important to try to <u>predict</u> eruptions, so the <u>damage</u> caused can be <u>minimised</u>.

Scientists can <u>monitor</u> the tell-tale signs that come before a <u>volcanic eruption</u>. Things such as <u>tiny earthquakes</u>, <u>escaping gas</u>, and changes in the <u>shape</u> of the volcano (e.g. <u>bulges</u> in the land where magma has built up under it) all mean an eruption is <u>likely</u>.

Uh oh...

Studying volcanoes — what a blast...

There are <u>different types</u> of volcano, but they've all got one thing in common — <u>throwing up lava</u> without bothering to rush to the toilet. You need to know the <u>characteristics</u> of the different types, as well as what's done to <u>predict</u> them.

Volcanic Eruption — Case Study

It's that time again, yep, <u>case study</u> time. Cram your memory full of these facts (the <u>cause</u> of the eruption, <u>impacts</u> and <u>responses</u> — don't think just learning the name and date will cut it).

The Soufrière Hills Volcano in Montserrat Erupted in 1997

Montserrat is a small island in the Caribbean Sea.

Montserrat

Caribbean sea

Date of eruption: <u>June 25th 1997</u> (small eruptions started in <u>July 1995</u>).

Size of eruption: <u>Large</u> — <u>4-5 million m³</u> of rocks and gas released.

Death toll: <u>19 killed</u>

Cause:
1) Montserrat is above a <u>destructive plate margin</u>, where the <u>Atlantic plate</u> is being <u>forced under</u> the <u>Caribbean plate</u>.
2) <u>Magma</u> rose up through <u>weak points</u> under the Soufrière hills forming an underground <u>pool of magma</u>.
3) The rock above the pool <u>collapsed</u>, opening a <u>vent</u> and causing the eruption.

You Need to Learn the Impacts and Responses

Primary impacts

1) <u>Large areas</u> were <u>covered</u> with <u>volcanic material</u> — the capital city <u>Plymouth</u> was buried under <u>12 m of mud and ash</u>.

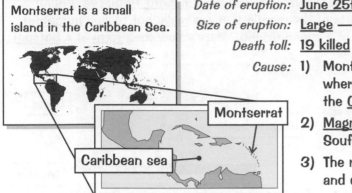
©istockphoto.com/Sean Hannah

2) Over <u>20 villages</u> and <u>two thirds of homes</u> on the island were <u>destroyed</u> by <u>pyroclastic flows</u> (fast-moving clouds of <u>super-heated gas</u> and <u>ash</u>).
3) <u>Schools</u>, <u>hospitals</u>, the <u>airport</u> and the <u>port</u> were <u>destroyed</u>.
4) <u>Vegetation</u> and <u>farmland</u> were <u>destroyed</u>.
5) <u>19 people died</u> and 7 were injured.

Secondary impacts

1) <u>Fires destroyed</u> many buildings including local <u>government offices</u>, the <u>police headquarters</u> and the town's central <u>petrol station</u>.
2) <u>Tourists stayed away</u> and <u>businesses</u> were <u>destroyed</u>, disrupting the <u>economy</u>.
3) <u>Population decline</u> — <u>8000</u> of the island's 12 000 inhabitants <u>have left</u> since the eruptions began in <u>1995</u>.
4) <u>Volcanic ash</u> from the eruption has <u>improved soil fertility</u>.
5) <u>Tourism</u> on the island is now <u>increasing</u> as people come to <u>see the volcano</u>.

Immediate responses

1) <u>People</u> were <u>evacuated</u> from the south to <u>safe areas</u> in the north.
2) <u>Shelters</u> were <u>built</u> to house evacuees.
3) Temporary <u>infrastructure</u> was also built, e.g. <u>roads</u> and <u>electricity supplies</u>.
4) The <u>UK</u> provided <u>£17 million</u> of <u>emergency aid</u> (Montserrat's an overseas territory of the UK).
5) <u>Local emergency services</u> provided support units to <u>search</u> for and <u>rescue</u> survivors.

Long-term responses

1) A <u>risk map</u> was created and an <u>exclusion zone</u> is in place. The south of the island is <u>off-limits</u> while the volcano is <u>still active</u>.
2) The <u>UK</u> has provided <u>£41 million</u> to develop the north of the island — <u>new docks</u>, an <u>airport</u> and <u>houses</u> have been built in the north.
3) The <u>Montserrat Volcano Observatory</u> has been set up to try and <u>predict</u> future eruptions.

Learn the Eruption and without Interruption learn the Eruption Disruption...

The eruption on Montserrat <u>wasn't all bad</u>, well... it was mostly bad, but there were <u>some positive impacts</u> — the ash improved the soil fertility and lots of tourists now go to gawp at the volcano. Every cloud has a silver lining and all that.

Unit 1A — The Restless Earth

Supervolcanoes

Normal volcanoes just not disastrous enough for you? What you need is a <u>supervolcano</u>. The same lava and ash combo, but <u>global destruction</u> thrown in. It's a brand <u>new topic</u> so an <u>exam question's likely</u>.

Supervolcanoes are Massive Volcanoes

Supervolcanoes are <u>much bigger</u> than standard volcanoes. They develop in a <u>handful of places</u> around the globe — at <u>destructive plate margins</u> or over parts of the <u>mantle</u> that are <u>really hot</u> (called <u>hotspots</u>), e.g. <u>Yellowstone National Park</u> in the USA is on top of a supervolcano. Here's how they <u>form</u> at a <u>hotspot</u>:

1) Magma <u>rises up</u> through <u>cracks</u> in the crust to form a large <u>magma basin</u> below the surface. The <u>pressure</u> of the magma causes a circular <u>bulge</u> on the surface <u>several kilometres wide</u>.

2) The bulge eventually <u>cracks</u>, creating <u>vents</u> for <u>lava</u> to escape through. The lava <u>erupts</u> out of the vents causing <u>earthquakes</u> and sending up gigantic plumes of <u>ash</u> and <u>rock</u>.

3) As the magma basin <u>empties</u>, the <u>bulge</u> is <u>no longer supported</u> so it <u>collapses</u> — spewing up <u>more lava</u>.

4) When the eruption's <u>finished</u> there's a <u>big crater</u> (called a <u>caldera</u>) left where the <u>bulge collapsed</u>. Sometimes these get filled with <u>water</u> to form a <u>large lake</u>, e.g. <u>Lake Toba</u> in Indonesia.

Before: Bulge · Crust · Magma basin · Mantle

During: Vents

After: Caldera · Collapsed basin

You need to know the <u>characteristics</u> of a supervolcano:

- <u>Flat</u> (unlike normal volcanoes, which are <u>mountains</u>).
- <u>Cover a large area</u> (much <u>bigger</u> than normal volcanoes).
- <u>Have a caldera</u> (unlike normal volcanoes where there's just a <u>crater</u> at the top).

When a Supervolcano Erupts there will be Global Consequences

Fortunately there are <u>only a few supervolcanoes</u> and an eruption <u>hasn't happened</u> for <u>tens of thousands of years</u>, e.g. the last one to erupt was the Lake Toba supervolcano <u>74 000 years ago</u>. When there is an eruption though, it's predicted that an <u>enormous area</u> will be <u>affected</u>:

SIMON TERREY / SCIENCE PHOTO LIBRARY

The predicted plume of ash from a supervolcanic eruption in Yellowstone National Park.

1) A supervolcanic eruption will throw out <u>thousands of cubic kilometres</u> of <u>rock</u>, <u>ash</u> and <u>lava</u> (much more than normal volcanoes, which usually produce a couple of cubic kilometres).

2) A thick <u>cloud</u> of <u>super-heated gas</u> and <u>ash</u> will flow at <u>high speed</u> from the volcano, <u>killing</u>, <u>burning</u> and <u>burying</u> everything it touches. Everything within <u>tens of miles</u> will be <u>destroyed</u>.

3) <u>Ash</u> will shoot <u>kilometres</u> into the air and <u>block out</u> almost all <u>daylight</u> over whole <u>continents</u>. This can <u>trigger mini ice ages</u> as less heat energy from the sun gets to Earth.

4) The <u>ash</u> will also <u>settle</u> over <u>hundreds of square kilometres</u>, <u>burying fields</u> and <u>buildings</u> (ash from normal volcanoes usually covers a couple of square kilometres).

The apocalypse is nigh — run away from the flat areas that look like a caldera...

Supervolcanoes are very different beasts from dome or shield volcanoes. Learn <u>how</u> they're formed and what the <u>predicted impact</u> will be. Oh, and you may want to reconsider your decision to move to Yellowstone National Park.

Earthquakes — Cause and Measurement

Plates can <u>get stuck</u> against each other (stupid plates) — when they become <u>unstuck</u> you get <u>earthquakes</u>.

Earthquakes Occur at All Three Types of Plate Margin

1) Earthquakes are caused by the <u>tension</u> that builds up at <u>all three</u> types of <u>plate margin</u>:

<u>Destructive margins</u> — <u>tension builds up</u> when one plate gets <u>stuck</u> as it's moving down past the other into the mantle.

<u>Constructive margins</u> — tension builds along <u>cracks within the plates</u> as they move away from each other.

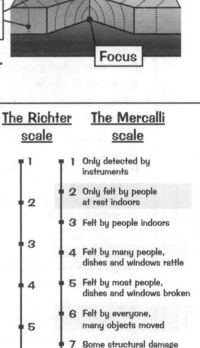

KEY
⁘ Earthquakes
∣ Plate margin

See p. 3 for more on plate margins.

<u>Conservative margins</u> — tension builds up when plates that are grinding past each other get <u>stuck</u>.

2) The plates eventually <u>jerk past each other</u>, sending out <u>shock waves</u> (vibrations). These vibrations are the <u>earthquake</u>.

3) The shock waves <u>spread out</u> from the <u>focus</u> — the point <u>in</u> the Earth where the earthquake <u>starts</u>. Near the focus the waves are <u>stronger</u> and cause <u>more damage</u>.

4) The <u>epicentre</u> is the point <u>on the Earth's surface</u> <u>straight above</u> the <u>focus</u>.

5) <u>Weak</u> earthquakes happen <u>quite often</u>, but <u>strong</u> earthquakes are <u>rare</u>.

Plates
Epicentre
Shock waves
Focus

Earthquakes can be Measured

Earthquakes can be <u>measured</u> using <u>two different scales</u>:

1 The <u>Richter scale</u>:

1) This measures the amount of <u>energy released</u> by an earthquake (called the <u>magnitude</u>).

2) Magnitude is measured using a <u>seismometer</u> — a machine with an arm that moves with the vibrations of the earth.

Seismometer reading

3) The Richter scale doesn't have an upper limit and it's <u>logarithmic</u> — this means that an earthquake with a magnitude of 5 is <u>ten times more powerful</u> than one with a magnitude of 4.

4) Most people <u>don't feel</u> earthquakes of magnitude <u>1-2</u>. <u>Major</u> earthquakes are <u>above 5</u>.

2 The <u>Mercalli scale</u>:

1) This measures the <u>effects</u> of an earthquake.

2) Effects are measured by asking eye witnesses for <u>observations</u> of what happened. Observations can be in the form of <u>words</u> or <u>photos</u>.

3) It's a scale of <u>1 to 12</u>.

The Richter scale	The Mercalli scale	
1	1	Only detected by instruments
2	2	Only felt by people at rest indoors
	3	Felt by people indoors
3	4	Felt by many people, dishes and windows rattle
4	5	Felt by most people, dishes and windows broken
5	6	Felt by everyone, many objects moved
	7	Some structural damage
6	8	Heavy structural damage
7	9	Massive structural damage, some buildings destroyed
	10	All buildings damaged, many destroyed
8	11	Most buildings destroyed
9+	12	Total destruction

Earthquakes are such jerks...

Boy, it's been a pretty destructive topic so far. It's earthquakes this time — you need to know what <u>causes</u> them at different <u>plate boundaries</u> and the different ways they can be <u>measured</u> — the <u>Richter</u> and <u>Mercalli</u> scales. Rock 'n roll.

Earthquakes — Case Studies

And you thought I'd forgotten all about the case studies. Shame on you.

Rich and Poor Parts of the World are Affected Differently

The effects of earthquakes and the responses to them are different in different parts of the world. A lot depends on how wealthy the part of the world is. I'd bet my budgie they'll want you to compare two earthquakes in a rich and a poor part of the world in the exam.

Earthquake in a rich part of the world:

Place: L'Aquila, Italy
Date: 6th April, 2009
Size: 6.3 on the Richter Scale
Cause: Movement along a crack in the plate at a destructive margin.

Earthquake in a poor part of the world:

Place: Kashmir, Pakistan
Date: 8th October, 2005
Size: 7.6 on the Richter Scale
Cause: Movement along a crack in the plate at a destructive margin.

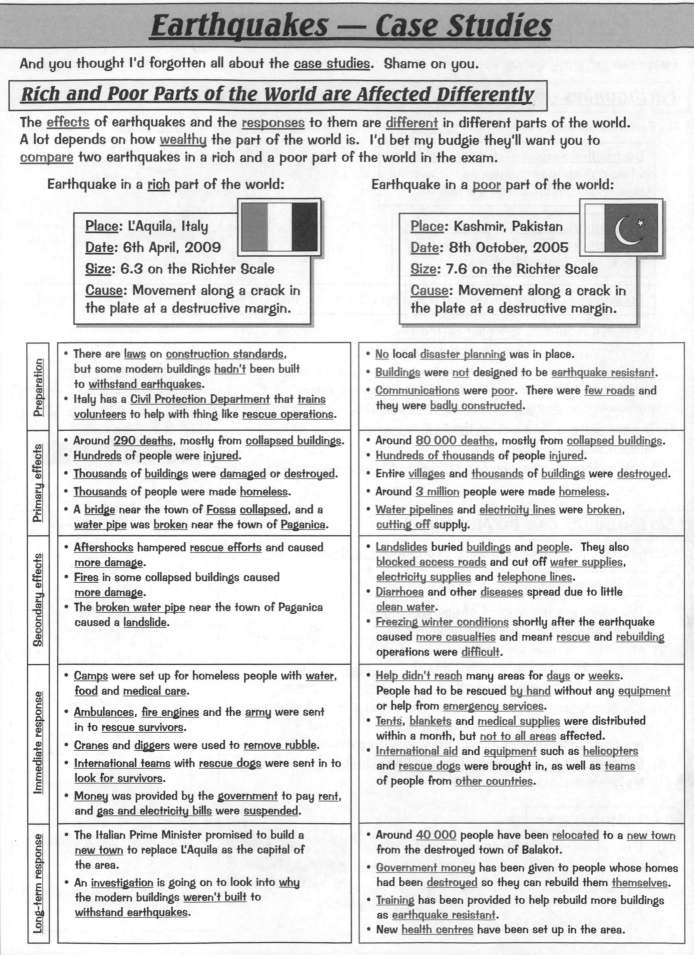

Preparation	• There are laws on construction standards, but some modern buildings hadn't been built to withstand earthquakes. • Italy has a Civil Protection Department that trains volunteers to help with thing like rescue operations.	• No local disaster planning was in place. • Buildings were not designed to be earthquake resistant. • Communications were poor. There were few roads and they were badly constructed.
Primary effects	• Around 290 deaths, mostly from collapsed buildings. • Hundreds of people were injured. • Thousands of buildings were damaged or destroyed. • Thousands of people were made homeless. • A bridge near the town of Fossa collapsed, and a water pipe was broken near the town of Paganica.	• Around 80 000 deaths, mostly from collapsed buildings. • Hundreds of thousands of people injured. • Entire villages and thousands of buildings were destroyed. • Around 3 million people were made homeless. • Water pipelines and electricity lines were broken, cutting off supply.
Secondary effects	• Aftershocks hampered rescue efforts and caused more damage. • Fires in some collapsed buildings caused more damage. • The broken water pipe near the town of Paganica caused a landslide.	• Landslides buried buildings and people. They also blocked access roads and cut off water supplies, electricity supplies and telephone lines. • Diarrhoea and other diseases spread due to little clean water. • Freezing winter conditions shortly after the earthquake caused more casualties and meant rescue and rebuilding operations were difficult.
Immediate response	• Camps were set up for homeless people with water, food and medical care. • Ambulances, fire engines and the army were sent in to rescue survivors. • Cranes and diggers were used to remove rubble. • International teams with rescue dogs were sent in to look for survivors. • Money was provided by the government to pay rent, and gas and electricity bills were suspended.	• Help didn't reach many areas for days or weeks. People had to be rescued by hand without any equipment or help from emergency services. • Tents, blankets and medical supplies were distributed within a month, but not to all areas affected. • International aid and equipment such as helicopters and rescue dogs were brought in, as well as teams of people from other countries.
Long-term response	• The Italian Prime Minister promised to build a new town to replace L'Aquila as the capital of the area. • An investigation is going on to look into why the modern buildings weren't built to withstand earthquakes.	• Around 40 000 people have been relocated to a new town from the destroyed town of Balakot. • Government money has been given to people whose homes had been destroyed so they can rebuild them themselves. • Training has been provided to help rebuild more buildings as earthquake resistant. • New health centres have been set up in the area.

Run away, run away — my immediate response to earthquakes...

The amount of damage an earthquake does, and the number of people that get hurt is different in different parts of the world. You need to know a couple of good examples, one from a rich part of the world and one from a poor part.

Tsunamis — Case Study

As if volcanoes and earthquakes weren't bad enough, if they happen out at sea they can cause tsunamis. A tsunami is a series of enormous waves caused when huge amounts of water get displaced.

An Earthquake caused a Tsunami in the Indian Ocean in 2004

1) There's a destructive plate margin along the west coast of Indonesia in the Indian Ocean.

2) On 26th December 2004 there was an earthquake off the west coast of the island of Sumatra measuring around 9.1 on the Richter scale.

3) The plate that's moving down into the mantle cracked and moved very quickly, which caused a lot of water to be displaced. This triggered a tsunami with waves up to 30 m high.

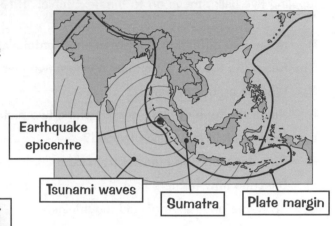

Earthquake epicentre

Tsunami waves

Sumatra

Plate margin

The Tsunami Affected Many Countries

The Indian Ocean tsunami was one of the most destructive natural disasters that's ever happened. It affected most countries bordering the Indian Ocean, e.g. Indonesia, Thailand, India and Sri Lanka. The effects of the tsunami were so bad because there was no early warning system:

1) Around 230 000 people were killed or are still missing.

2) Whole towns and villages were destroyed — over 1.7 million people lost their homes.

3) The infrastructure (things like the roads, water pipes and electricity lines) of many countries was severely damaged.

4) 5-6 million people needed emergency food, water and medical supplies.

Sri Lankan coastline before the tsunami

Sri Lankan coastline during the tsunami

5) There was massive economic damage. Millions of fishermen lost their livelihoods, and the tourism industry suffered because of the destruction and because people were afraid to go on holiday there.

6) There was massive environmental damage. Salt from the seawater has meant plants can't grow in many areas. Mangroves, coral reefs, forests and sand dunes were also destroyed by the waves.

The Response Involved a lot of International Aid

Short-term responses:

1) Within days hundreds of millions of pounds had been pledged by foreign governments, charities, individuals and businesses to give survivors access to food, water, shelter and medical attention.

2) Foreign countries sent ships, planes, soldiers and teams of specialists to help rescue people, distribute food and water and begin clearing up.

Long-term responses:

1) Billions of pounds have been pledged to help re-build the infrastructure of the countries affected.

2) As well as money, programmes have been set up to re-build houses and help people get back to work.

3) A tsunami warning system has been put in place in the Indian Ocean.

4) Disaster management plans have been put in place in some countries. Volunteers have been trained so that local people know what to do if a tsunami happens again.

Sorry dudes, but you don't want to surf this one...

Crikey, tsunamis can wreak as much havoc as the earthquakes or volcanoes that cause them. You need to know a case study of a tsunami and it might as well be this one — learn what caused it, the effects it had and the responses to it.

Revision Summary for The Restless Earth

What a cracking topic, earth-shattering one might even say. It may be a topic about disasters, but get a load of these questions down you and there won't be any kind of disaster in the exam. I know it looks like there's a lot of stuff here, but you'll be surprised how much you just learnt. Try them out a few at a time, then check the answers on the pages. Once you can answer them all standing on your head and juggling five balls, move on to the next topic. It rocks, trust me...

1) Name two differences between continental plates and oceanic plates.

2) Name the type of plate margin where two plates are moving towards each other.

3) What is an ocean trench?

4) Name the type of plate margin where two plates are moving sideways against each other.

5) How are fold mountains formed?

6) Give one way that humans use fold mountain areas.

7) Describe how farmers have adapted to the steep slopes in fold mountain areas.

8) a) Name one range of fold mountains.

    b) Describe three ways humans use the area.

    c) Describe how people have adapted to the conditions of the area.

9) What's magma called when it erupts onto the surface?

10) Which type of volcano is made up of layers of ash and lava? Name an example.

11) Which type of volcano is formed when the lava is runny? Name an example.

12) Give one thing that scientists do to try and predict a volcanic eruption.

13) a) Name a volcanic eruption and state when and where it happened.

    b) Describe two negative primary impacts of the eruption.

    c) Describe two negative secondary impacts of the eruption.

    d) Give two positive impacts of the eruption.

    e) Give two immediate responses.

    f) What were two of the long-term responses?

14) Where do supervolcanoes form?

15) Give one way that a supervolcanic eruption is different from a volcanic eruption.

16) Give one predicted effect of a supervolcanic eruption.

17) What causes earthquakes?

18) What's the point in the Earth called where an earthquake starts?

19) What's the name of the point on the surface of the Earth above where an earthquake starts?

20) What does the Richter scale measure?

21) a) Give an example of an earthquake in a rich part of the world.

    b) Describe two effects of the earthquake and two responses to it.

22) a) Give an example of an earthquake in a poor part of the world.

    b) Describe two effects of the earthquake and two responses to it.

23) What causes a tsunami?

24) a) Give an example of a tsunami.

    b) Describe two effects of the tsunami.

Types of Rock

There are <u>three types</u> of <u>rock</u> — hard, heavy and punk... erm, I mean <u>igneous</u>, <u>sedimentary</u> and <u>metamorphic</u>. Rock type depends on how the rock was <u>formed</u>.

Igneous Rocks are Formed from Magma that's Cooled Down

<u>All</u> igneous rocks are formed when <u>molten rock</u> (magma) from the mantle <u>cools down</u> and <u>hardens</u>. There are <u>two types</u> depending on <u>where</u> the magma has cooled down:

The mantle is a layer of molten rock deep in the Earth.

1 INTRUSIVE igneous rocks, e.g. <u>granite</u>

1) These form when magma cools down <u>below</u> the Earth's surface.

2) The magma <u>cools down</u> very <u>slowly</u>, forming <u>large crystals</u> that give the rocks a <u>coarse texture</u>.

3) Large <u>domes</u> of cooled magma form domes of igneous rock called <u>batholiths</u>.

4) Where the magma has flowed into <u>gaps</u> in the surrounding rock it forms <u>dykes</u> (in <u>vertical</u> gaps) and <u>sills</u> (in <u>horizontal</u> gaps).

2 EXTRUSIVE igneous rocks, e.g. <u>basalt</u>

1) These form when magma cools down <u>after</u> it's <u>erupted from</u> a <u>volcano</u> onto the Earth's <u>surface</u>.

2) The magma <u>cools down</u> very <u>quickly</u>, forming <u>small crystals</u> that give the rocks a <u>fine texture</u>.

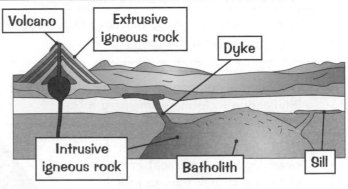

Volcano — Extrusive igneous rock — Dyke — Intrusive igneous rock — Batholith — Sill

Sedimentary Rocks are Formed from Compacted Sediment

Sedimentary rocks are formed when layers of <u>sediment</u> are <u>compacted together</u> until they become <u>solid rock</u>. The process of compaction is called <u>lithification</u>. Here are a couple of examples:

1) <u>Carboniferous limestone</u> and <u>chalk</u> are formed from <u>calcium carbonate</u>. Layers of <u>tiny shells</u> and <u>skeletons</u> of dead sea creatures are deposited on the <u>sea bed</u> and <u>compacted together</u> over time.

2) <u>Clays</u> and <u>shales</u> are made from <u>mud</u> and <u>clay minerals</u>. The particles have been <u>eroded</u> from older rocks, deposited in <u>layers</u> on lake or sea beds then compacted together.

Sedimentary rocks often contain fossils.

Metamorphic Rocks are Formed by Heat and Pressure

<u>Metamorphic</u> rocks are formed when other rocks (igneous, sedimentary or older metamorphic rocks) are <u>changed</u> by <u>heat</u> and <u>pressure</u>:

1) Rocks <u>deep</u> in the Earth are <u>changed</u> by the <u>pressure</u> from the <u>weight</u> of the <u>material above them</u>.

2) When <u>tectonic plates collide</u>, rocks are <u>changed</u> by the massive <u>heat</u> and <u>pressure</u> that <u>builds up</u>.

3) <u>Magma</u> from the mantle <u>heats</u> the rocks in the crust, causing them to <u>change</u>.

The new rocks are <u>harder</u> and <u>more compact</u>, e.g. limestone becomes <u>marble</u> and clay becomes <u>slate</u>.

You need to know the <u>location</u> of the three rock types in the <u>UK</u>.

■ Igneous rocks
■ Sedimentary rocks
□ Metamorphic rocks

Stuck between a rock and a hard exam...

<u>Don't skip this page</u> 'cos it's easy to get the different rocks <u>mixed up</u>. Use the names to help you remember how they're formed — sedimentary is from <u>sediment</u>, metamorphic is rock that's <u>morphed</u> (changed) and <u>igneous</u> is, well, <u>the other one</u>.

The Rock Cycle

You might think <u>once a rock exists</u> then <u>that's it</u> — but weirdly they can <u>change</u> from <u>one type into another</u>.

The Formation of all Rock Types is Linked by the Rock Cycle

The <u>rock cycle</u> shows how igneous, sedimentary and metamorphic rocks are <u>formed</u>, and <u>how</u> one type is <u>changed</u> into another:

1) <u>Weathering</u> (the <u>breakdown</u> of rocks) of all three rock types creates loose <u>sediment</u>.

2) This makes it <u>easier</u> for <u>erosion</u> (the <u>removal</u> of rock) to occur.

3) The sediment is <u>transported away</u> (e.g. by rivers) and <u>deposited</u> on the <u>sea bed</u>.

4) Sediment is <u>compacted</u> on the sea bed through <u>lithification</u> to form <u>sedimentary rocks</u>.

5) <u>Heat</u> and <u>pressure</u> (e.g. from overlying layers of rock) can change any rock type to new <u>metamorphic rock</u>.

6) <u>Melting</u> of any rock type (e.g. in the mantle) <u>creates</u> <u>magma</u>. When magma <u>cools</u>, <u>igneous rocks</u> are formed.

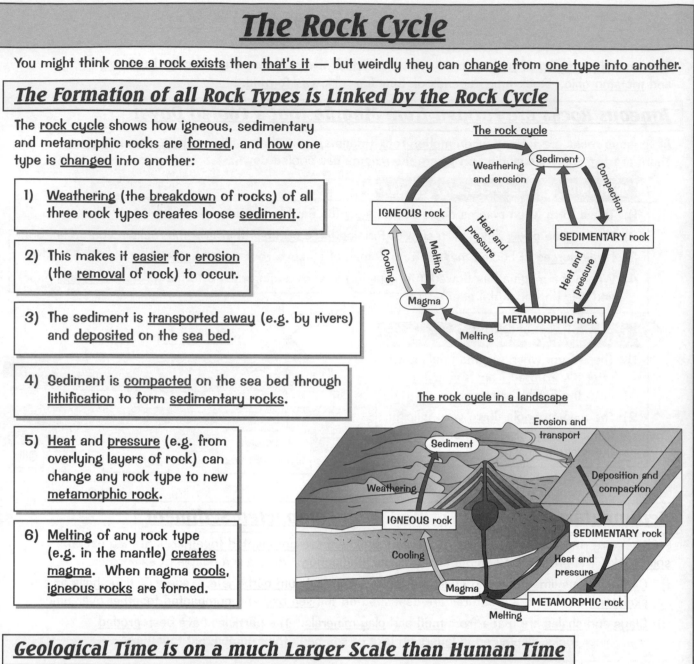

The rock cycle

The rock cycle in a landscape

Geological Time is on a much Larger Scale than Human Time

Geological period	Began, million years before present
Quaternary	2.6
Tertiary	65
Cretaceous	145
Jurassic	215
Triassic	245
Permian	285
Carboniferous	360
Devonian	410
Silurian	440
Ordovician	505
Cambrian	585

Humans evolved

Chalk formed in the UK

Clay formed in the UK

Carboniferous limestone formed in the UK

Granite formed in the UK

These are the major periods when each of these rocks formed — some of them formed at other times too.

It takes a <u>very long time</u> for rocks to <u>form</u> and go through the rock cycle.

1) The table on the left shows all the <u>most recent geological periods</u> (you don't need to learn their names).

2) The rocks found in the UK <u>today</u> were all formed <u>many millions</u> of years ago.

3) <u>Humans</u> (Homo sapiens) have only been around for the last <u>200 000 years</u>, so the rock cycle is on a totally <u>different time scale</u> to <u>human time</u>.

Rock cycles — really heavy and uncomfortable to sit on...

I feel pretty small and a lot less worried about how old I'll be on my next birthday now — these <u>rocks</u> really are <u>old codgers</u>. Make sure you can scribble down the <u>rock cycle</u> and explain how one rock type can be changed into another.

Weathering

Weathering is the <u>breakdown</u> of rocks <u>where they are</u> (the material created doesn't get taken away like with erosion). There are <u>three</u> types — <u>mechanical</u>, <u>chemical</u> and <u>biological</u>.

Mechanical Weathering — Rocks are Broken Down by Physical Processes

Mechanical weathering is the breakdown of rocks <u>without changing</u> their <u>chemical composition</u>.
You need to know about <u>two types</u> of mechanical weathering:

FREEZE-THAW weathering

1) In some areas (e.g. upland Britain in winter), the temperature is <u>above 0 °C</u> during the <u>day</u>, and <u>below 0 °C</u> at <u>night</u>.

2) During the day, water <u>gets into cracks in rocks</u>, e.g. granite.

3) At night, the water <u>freezes</u> and <u>expands</u>, which puts <u>pressure</u> on the rock.

4) The water <u>thaws</u> the next day, releasing the pressure, then <u>refreezes</u> the next night.

5) <u>Repeated freezing</u> and <u>thawing</u> <u>widens</u> the <u>cracks</u> and causes the rock to <u>break up</u>.

EXFOLIATION weathering

1) Some areas have a <u>big daily temperature range</u>, e.g. deserts can be 40 °C in the day and 5 °C at night.

2) Each day the <u>surface layers</u> of rock <u>heat up</u> and <u>expand faster</u> than the <u>inner layers</u>.

3) At night the surface layers <u>cool down</u> and <u>contract faster</u> than the inner layers

4) This <u>creates pressure within the rock</u> and causes thin <u>surface layers</u> to <u>peel off</u>.

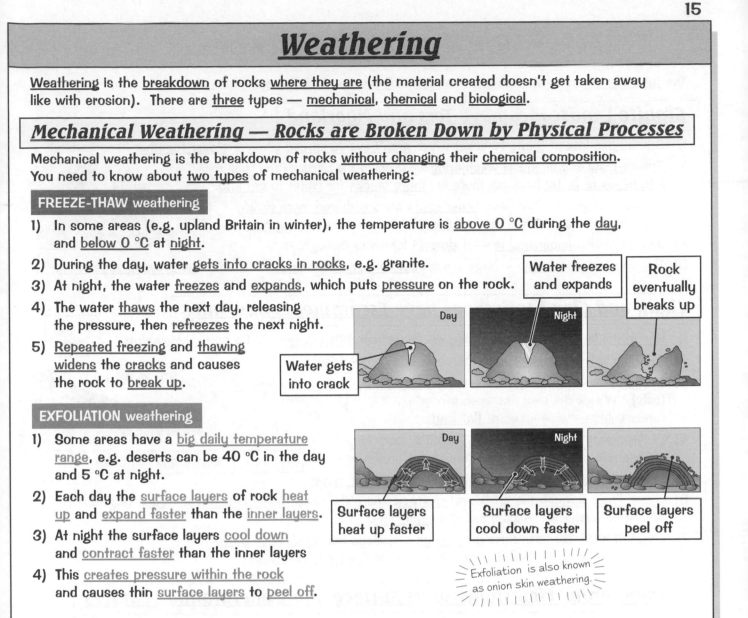

Water freezes and expands

Rock eventually breaks up

Day   Night

Water gets into crack

Surface layers heat up faster

Surface layers cool down faster

Surface layers peel off

Exfoliation is also known as onion skin weathering.

Chemical Weathering — Rocks are Broken Down by Being Dissolved

Chemical weathering is the breakdown of rocks by <u>changing</u> their <u>chemical composition</u>.
You need to know about <u>two types</u> of chemical weathering:

SOLUTION weathering

1) Some <u>minerals</u> that make up rocks are <u>soluble in water</u>, e.g. rock salt.

2) The minerals <u>dissolve</u> in rainwater, breaking the rock down.

CARBONATION weathering

1) Rainwater has <u>carbon dioxide</u> dissolved in it, which makes it a <u>weak carbonic acid</u>.

2) Carbonic acid <u>reacts</u> with rocks that contain <u>calcium carbonate</u>, e.g. carboniferous limestone, so the rocks are <u>dissolved</u> by the rainwater.

Biological Weathering — Rocks are Broken Down by Plants and Animals

Biological weathering is the breakdown of rocks by <u>living things</u>:

1) <u>Plant roots</u> break down rocks by <u>growing into cracks</u> on their surfaces and <u>pushing them apart</u>.

2) <u>Burrowing animals</u> may <u>loosen</u> small amounts of rock material.

This weathering stuff cracks me up...

So, <u>exfoliating</u> isn't just something that happens at a beauty parlour. Rocks need younger looking, more plumped up skin too. Learn all about the <u>three types</u> of <u>weathering</u> and you won't break down in the exam.

Rocks and Landscapes

The <u>type of rock</u> in an area affects the <u>type</u> of <u>landscape that forms</u>.

Granite Landscapes have Tors and Moorland

1) Granite has lots of <u>joints</u> (cracks) which <u>aren't evenly spread</u> (they're <u>closer together</u> in some bits).

2) <u>Freeze-thaw</u> and <u>chemical weathering</u> wear down the parts of the rock with <u>lots of joints</u> <u>faster</u> because there are <u>more cracks</u> for <u>water</u> to <u>get into</u>.

3) Sections of granite that have <u>fewer joints</u> are <u>weathered more slowly</u> than the surrounding rock and <u>stick out</u> at the surface forming <u>tors</u>.

Tors

4) Granite is also <u>impermeable</u> — it <u>doesn't</u> let water through.

5) This creates <u>moorlands</u> — large areas of <u>waterlogged</u> and <u>acidic</u> soil, with <u>low-growing vegetation</u>.

Chalk and Clay Landscapes have Escarpments and Vales

1) <u>Horizontal layers</u> of <u>chalk</u> and <u>clay</u> are sometimes <u>tilted diagonally</u> by <u>earth movements</u>.

2) The <u>clay</u> is <u>less resistant</u> than the chalk so is <u>eroded faster</u>.

3) The <u>chalk</u> is left sticking out forming <u>escarpments</u> (hills). Where the clay has been eroded it forms <u>vales</u> — wide areas of <u>flat land</u>.

4) Escarpments have <u>steep slopes</u> (called a scarp slope) on one end, and <u>gentle slopes</u> (dip slope) on the other.

5) Chalk is an <u>aquifer</u> — a <u>permeable</u> rock that <u>stores water</u>.

6) Water <u>flows through</u> the chalk and <u>emerges</u> where the chalk <u>meets impermeable rock</u> (e.g. clay). Where the water emerges is called a <u>spring line</u>.

7) Areas of chalk can also have <u>dry valleys</u> — valleys that <u>don't</u> have a <u>river</u> or stream flowing in them because the water is <u>flowing underground</u>.

Spring line · **Dry valley** · **Escarpment** · **Aquifer** · **Clay** · **Vale** · **Chalk** · **Daisy**

Escarpments are also called cuestas.

Carboniferous Limestone forms Surface and Underground Features

Rainwater slowly <u>eats away</u> at carboniferous limestone through <u>carbonation weathering</u> (see p. 15). Most weathering happens along <u>joints</u> in the rock, creating some spectacular <u>features</u>:

1) <u>Limestone pavements</u> are flat areas of limestone with <u>blocks</u> separated by <u>weathered-down joints</u>.

2) <u>Swallow holes</u> are <u>weathered holes</u> in the surface.

3) <u>Caverns</u> form beneath swallow holes where the limestone has been <u>deeply weathered</u>.

4) <u>Limestone gorges</u> are steep sided gorges formed when <u>caverns collapse</u>.

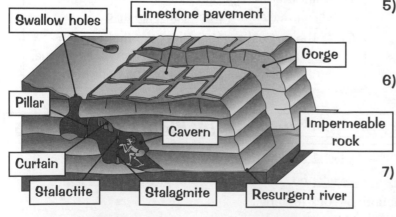

Swallow holes · **Limestone pavement** · **Gorge** · **Pillar** · **Impermeable rock** · **Cavern** · **Curtain** · **Stalactite** · **Stalagmite** · **Resurgent river**

5) Limestone is <u>permeable</u>, so limestone areas also have <u>dry valleys</u> and <u>resurgent rivers</u> (rivers that pop out at the surface when limestone is on top of impermeable rock).

6) Water seeping through limestone contains <u>dissolved minerals</u>. When the water drips into a cavern the minerals <u>solidify</u> and <u>build up</u> over time to produce <u>stalactites</u> (on the ceiling) and <u>stalagmites</u> (on the <u>ground</u>).

7) When stalagmites and stalactites <u>meet</u> in the middle they form a <u>pillar</u>. When water <u>flows</u> in as a <u>sheet</u> a <u>curtain</u> builds up.

Shut your swallow hole and get on with it...

There's a fair bit to learn on this page, I'll grant you that, but it's all pretty straightforward stuff. Scribble down the <u>diagrams</u> and <u>label</u> them, then make sure you can <u>explain</u> how <u>each landform</u> you've labelled <u>is made</u>.

Using Landscapes — Case Studies

You need to know how people use rock landscapes for resources, farming, tourism and water supplies.

Granite Areas are Used for Stone, Tourism, Farming and Water

1) Granite is quarried and used as a building stone for things like flooring and worktops.

2) Tourists are attracted to the features of granite landscapes, e.g. tors and moorland (see previous page).

3) There are opportunities for rearing livestock on granite areas, but the land isn't great for arable farming (growing crops) or dairy farming because the soils are acidic and waterlogged.

4) Granite's impermeable so granite areas are good places to build reservoirs.

Case Study — Dartmoor, Devon

1) Dartmoor has lots of granite quarries, e.g. Meldon Quarry. Granite from Dartmoor was used to build Nelson's Column.

2) Millions of people visit Dartmoor every year to enjoy the features of the granite landscape, e.g. Bowerman's Nose and Hound Tor are popular attractions.

3) In 2000, there were over 290 000 hectares of land used for rearing livestock (only 900 hectares of land were used for arable farming)

4) There are 8 reservoirs in Dartmoor, e.g. Burrator Reservoir supplies water to Plymouth.

Bowerman's Nose

Dartmoor

Chalk and Clay Areas are Used for Cement, Tourism, Farming and Water

1) Chalk is quarried and used to make cement, which is then used to make building materials like concrete.

2) Tourists are attracted to the features of chalk and clay landscapes, e.g. escarpments and vales (see previous page).

3) There are opportunities for arable farming, livestock rearing and dairy farming on chalk and clay areas (clay vales are wide, flat, grassy areas).

4) Chalk is an aquifer — a permeable rock that holds water. Aquifers are often used as a source of drinking water — water is taken out through wells and is also pumped out of the rocks.

Case Study — The Lincolnshire Wolds, Lincolnshire

1) In 2001, 438 000 tonnes of chalk were quarried from Lincolnshire.

2) The Wolds are an Area of Outstanding Natural Beauty — many tourists come to the Wolds for the scenery and activities like walking, e.g. along the Viking Way (a long-distance footpath).

3) Around 80% of the Wolds is used as farmland to grow crops.

4) There's a major chalk aquifer underneath the Wolds — it supplies water to Lincolnshire.

The Lincolnshire Wolds

Get off my footpath.

Granite — bad for crops, good for stylish kitchens...

And you thought rocks were just boring grey lumps that aren't very useful. Shame on you. Learn a few of the ways granite and chalk and clay landscapes are used and maybe the rocks will forgive you. Maybe, but don't count on it.

Using Landscapes — Case Studies

I don't know about you, but I'm getting a little bit rocked out. Never mind, you're over half way through this topic now. There's just limestone area uses, a bit of tourism and some quarrying fun to go.

Limestone Areas are Used for Stone, Cement, Tourism and Farming

1) Limestone is quarried and used as a building stone to make things like floors and walls, e.g. in churches.

2) Limestone is also used to make cement.

3) Tourists are attracted to the features of limestone landscapes, e.g. limestone pavements, gorges and caverns (see p. 16).

4) There are opportunities for dairy farming or rearing livestock on limestone areas. There are also opportunities for arable farming (growing crops), but in some places the soil is quite alkaline.

Case Study — The White Peak, Peak District

1) Tunstead Quarry near Buxton produces about 5.5 million tonnes of limestone every year.

2) The cement processing plant inside Tunstead Quarry makes about 800 000 tonnes of cement each year.

3) Thor's Cave is a limestone cavern in the Manifold Valley, Staffordshire — it's popular with cavers, walkers and climbers.

4) The White Peak is mainly used for intensive dairy farming.

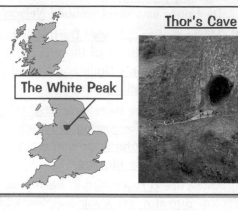

The White Peak

Thor's Cave

MARTIN BOND / SCIENCE PHOTO LIBRARY

Tourism in Any Area has Costs and Benefits

COSTS

- Large numbers of people can cause footpath erosion and littering around attractions.
- Lots of tourists cause traffic congestion.
- People may be attracted into the tourist industry from jobs like farming, causing a decline in traditional jobs.
- Many tourists own second homes in rural areas. They're absent through most of the year, so things like local shops close down.

BENEFITS

- Tourism creates jobs and brings money into the local economy.
- Farmers can diversify their business to get extra income from tourism, e.g. using buildings for camping barns or running bed and breakfasts.
- New businesses may be set up in the area to cater for the tourists, e.g. souvenir shops or hotels.

Case Study — Dartmoor, Devon

These tourists make me so congested...

COSTS

- Increased traffic means the narrow roads are getting congested. Grass verges are being damaged by tourists parking on them.
- Tourists disturb grazing animals on the moor — especially when dogs are let loose.

BENEFITS

- 4.5 million tourists visit Dartmoor every year. This creates around 3000 jobs.
- In 2003 tourism generated £120 million for the local economy.

Thor's Cave — I bet it took him ages to hammer it out...

Tourists, eh, everyone loves to hate 'em but they don't half bring a few benefits to some places. Learn about the uses of limestone areas and the benefits and costs of tourism and you win a free holiday to the White Peak*.

* No you don't, but learn this stuff anyway.

Quarrying Impacts — Case Study

People love rocks... and not just the shiny ones. No, seriously, they love them so much we have to have quarries just to meet the demand. Quarries can be quite quarrelsome things though...

Quarries Have Advantages and Disadvantages

A quarry is basically a massive pit in the ground that rock is taken from. Quarries bring advantages to some people and disadvantages to other people. This means people disagree about things like where they should go, or if we should have them at all. Here's a nice table so you can see both sides of the story:

	Advantages	Disadvantages
Economic	Quarries employ lots of local people — this brings more money into the local economy.	Tourists could be put off from visiting an area that has a quarry because they're noisy and they're an eyesore — this reduces the amount of money made from tourism.
Economic	When a quarry's built, good transport links are also built for the trucks that carry the stone — an improved infrastructure will attract other businesses and boost the local economy.	When a quarry is closed down it costs money to make them safe, e.g. by filling in holes and putting up warning signs.
Social	Some quarries are used by schools and colleges for educational visits.	People are annoyed by the heavy traffic caused by slow vehicles leaving quarries.
Social	Quarries that have been closed down can be used for recreation, e.g. for climbing.	Quarries are a very dangerous environment — people could be harmed or killed in them.
Environmental	The landscape is often restored after quarries are closed down — this can create new habitats and attract new species of wildlife to an area.	While they're operating the quarries have a massive impact on the environment — habitats are destroyed and the resources in the landscape are depleted.

Whatley Quarry is a Limestone Quarry in Somerset

Whatley Quarry is one of the largest quarries in the UK. It produces around 5 million tonnes of rock every year. You need to know about the economic, social and environmental advantages and disadvantages of Whatley Quarry:

Advantages

The quarry employs a lot of people — around 100 people work at the quarry full-time.

The quarry has a study centre — around 4000 people from schools and colleges visit every year.

The quarry has donated stone to build a cycle track in the local area.

Whatley Quarry

Disadvantages

Each blast removes around 25 000 tonnes of rock from the quarry — this creates a lot of noise and disrupts wildlife.

The quarry is around 1.5 km long and 0.6 km wide — it's destroyed a large area of habitat.

A man was killed in an industrial accident at the quarry in 2008.

Don't dig yourself into a hole with this lot...

I'd say one of the best things about quarries is that rock gets blown up — explosions are well cool. Don't listen to me though — you need to learn about the advantages and disadvantages of quarries, both in general and in the case study.

Quarrying Management — Case Study

My, my, aren't you lucky a person — another <u>case study</u> is waiting for you just below. It's another <u>quarry</u>, but this time it's about <u>sustainable management</u>... I won't keep you any longer. Have fun.

The Sustainable Management of Quarries is Important

1) Sustainable management is all about <u>meeting</u> the <u>needs</u> of people <u>today</u>, <u>without hindering</u> the ability of people in the <u>future</u> to <u>meet their own needs</u>.

2) It involves getting what we want <u>without damaging</u> or <u>altering</u> the <u>environment</u> in an <u>irreversible way</u>.

3) <u>Quarries</u> need sustainable management because they <u>could seriously damage</u> the <u>environment</u>, e.g. by <u>destroying habitats</u> and <u>local wildlife</u>.

Llynclys Quarry is a Limestone Quarry in Shropshire

Llynclys Quarry is a quarry in <u>Shropshire</u> that covers <u>65 hectares</u> of land. Some quarries are <u>abandoned</u> when the resources are <u>exhausted</u>, but Llynclys Quarry is being <u>sustainably managed</u> in areas where extraction has <u>finished</u>. This <u>minimises</u> the <u>environmental impact</u> of the quarry.

You need to know about some of the <u>sustainable management strategies</u> being used:

Sustainable management strategies

1) Areas of the quarry where work has finished are being <u>restored</u> to the <u>grassland</u>, <u>shrubland</u> and <u>woodland habitats</u> that <u>used to exist</u> before quarrying started. About <u>14%</u> of the quarry has been <u>restored</u> so far.

Llynclys Quarry

2) A <u>wetland habitat</u> has been <u>created</u> at the quarry. This encourages lots of different species to live in the area, e.g. the <u>insects</u> living in the wetland are a <u>food supply</u> for <u>bats</u>.

3) The habitats are <u>attracting animals</u> that <u>used to be in the area</u> before it was a quarry, e.g. the <u>Grizzled Skipper butterfly</u> has <u>returned</u> to the area.

New land uses in restored parts of the quarry

1) Some parts are used for <u>farming</u> — <u>sheep graze</u> around the <u>wetland</u>, which also <u>controls</u> the <u>growth of vegetation</u> there.

2) <u>Recreational activities</u>, e.g. <u>walking</u>, are <u>allowed</u> in the restored parts of the quarry.

3) <u>Tourism</u> has been <u>boosted</u> — the <u>restored habitats</u> and an <u>annual open day</u> are <u>attracting</u> a lot of <u>visitors</u> to the quarry.

Llynclys — send in your pronunciation suggestions on a postcard...

The hardest thing about this <u>case study</u> is saying the quarry's name, so you don't have any excuse for not learning all the different <u>sustainable management strategies</u>. Now this topic's done you can rock on to the revision summary...

Revision Summary for Rocks, Resources and Scenery

After the blast you've had with quarrying you'd expect a revision summary to be a bit a of let down. On the contrary, this will give you a chance to shine and show the world (or the people within earshot) just how much you know about the world of rocks. If you get stuck then have a flick back through the pages, but remember — don't move on to the next topic until you can answer each question with the elegance of a granite worktop.

1) a) How are igneous rocks formed?

    b) Name the two types of igneous rock.

2) What are sedimentary rocks formed from?

3) Describe how metamorphic rocks are formed.

4) What part does weathering play in the rock cycle?

5) What part does erosion play in the rock cycle?

6) Put these rock types in the order that they formed in the UK:
carboniferous limestone, chalk, granite, clay.

7) What is mechanical weathering?

8) a) Describe how freeze-thaw weathering takes place.

    b) Describe how exfoliation weathering takes place.

9) What is chemical weathering?

10) Name the two types of chemical weathering.

11) Give two examples of biological weathering.

12) Describe how tors form.

13) What are moorlands?

14) How do chalk escarpments form?

15) What is a vale?

16) What is an aquifer?

17) Where does a spring line form?

18) Give two surface features of a carboniferous limestone landscape.

19) What is a resurgent river?

20) How do stalactites and stalagmites form?

21) a) Name a granite area.

    b) Give the main type of farming carried out in the area.

22) a) Name a limestone area.

    b) Give two uses of the limestone produced in the area.

23) a) Give one economic disadvantage of quarries.

    b) Give one social advantage of quarries.

24) a) Give one economic advantage of a named quarry.

    b) Give one environmental disadvantage of the same quarry.

25) a) What is sustainable management?

    b) Why is the sustainable management of quarries important?

26) a) Give one example of sustainable management during the extraction of rock at a named quarry.

    b) Give two examples of sustainable management after the extraction of rock at a named quarry.

UK Climate

You may think it <u>rains</u> a lot in the <u>UK</u> (and you'd be right)... Well, now's your chance to find out <u>why</u>.

The UK has a Mild Climate — Cool, Wet Winters and Warm, Wet Summers

Temperature
Follows a <u>seasonal</u> pattern.
<u>Highest</u>: Jul-Aug (average 19 °C).
<u>Lowest</u>: Jan-Feb (average 6 °C).
<u>Temperature range</u>: <u>13 °C</u>.

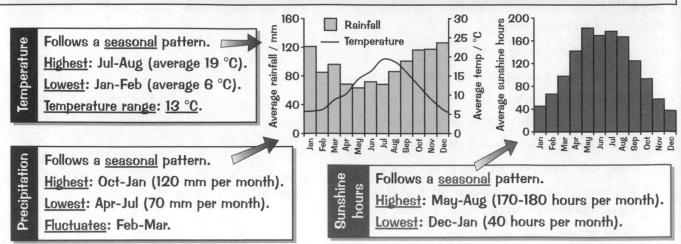

Precipitation
Follows a <u>seasonal</u> pattern.
<u>Highest</u>: Oct-Jan (120 mm per month).
<u>Lowest</u>: Apr-Jul (70 mm per month).
<u>Fluctuates</u>: Feb-Mar.

Sunshine hours
Follows a <u>seasonal</u> pattern.
<u>Highest</u>: May-Aug (170-180 hours per month).
<u>Lowest</u>: Dec-Jan (40 hours per month).

There are Five Main Reasons Why the Climate Varies Within the UK

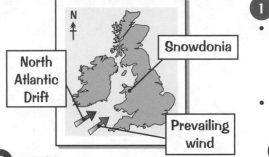

North Atlantic Drift

Snowdonia

Prevailing wind

1 <u>LATITUDE</u> (how far north or south of the equator a place is)
- The <u>higher</u> in latitude you go, the <u>colder</u> it gets. The <u>Sun</u> is at a <u>lower angle</u> in the sky, so its <u>heat energy</u> is <u>spread over more</u> of the Earth's surface — each place receives <u>less heat energy</u> than at <u>lower latitudes</u>.
- <u>Southern</u> parts of the UK are <u>warmer</u> than <u>northern</u> parts because of their <u>lower latitude</u>.

2 <u>WINDS</u>
- The UK's <u>most common</u> (<u>prevailing</u>) winds are <u>from</u> the <u>south west</u>. They bring <u>warm</u>, <u>moist air</u>, which makes the UK <u>warm</u> and <u>wet</u>.
- The <u>west</u> of the UK gets <u>more</u> of the <u>warmth</u> and <u>rain</u> than the <u>east</u> because the winds come from the south west.

3 <u>DISTANCE FROM THE SEA</u>
- Areas <u>near</u> the sea are <u>warmer</u> than inland areas in <u>winter</u> because the <u>sea stores</u> up <u>heat</u> and <u>warms the land</u>.
- Areas <u>near</u> the sea are <u>cooler</u> in <u>summer</u> because the <u>sea</u> takes a <u>long time</u> to <u>heat up</u> and so <u>cools the land down</u>.
- The <u>west</u> of the UK gets <u>warmed more</u> than the <u>east</u> because of a <u>warm ocean current</u> coming from the <u>south west</u> called the <u>North Atlantic Drift</u>.

4 <u>PRESSURE</u>
- <u>Low pressure</u> weather systems have <u>lots of rainfall</u> because the air is <u>rising</u> and <u>water vapour</u> is <u>condensing</u>. <u>High pressure</u> systems have <u>dry</u> <u>weather</u> because the air is <u>falling</u>.
- Low pressure weather systems come from the <u>west</u>, so the west of the UK is <u>wetter</u>.

5 <u>ALTITUDE</u> (how high the land is)
- The <u>higher up</u> you go the <u>colder</u> it gets because the air is <u>thinner</u> so <u>less heat</u> energy is <u>trapped</u>.
- Higher areas get <u>more rainfall</u> as air is <u>forced upwards</u> and the <u>water vapour condenses</u> into rain clouds.
- So <u>high altitude</u> parts of the UK (e.g. Snowdonia) are <u>colder</u> and <u>wetter</u> than low altitude areas.

Some of these factors also <u>explain</u> why the climate of the <u>whole</u> of the <u>UK</u> is <u>mild</u>:
- It's <u>not really hot</u> or <u>really cold</u> because it's a <u>mid-latitude</u> country, and it has the <u>North Atlantic Drift</u>.
- The UK has <u>both dry</u> and <u>rainy weather</u> because it gets both <u>high</u> and <u>low pressure</u> weather systems.

Warm and wet — they got the wet bit right...

There's quite a lot to remember about why the climate varies in the UK, but it'll be worth your while — they like to ask questions about <u>why two places</u> in the UK have <u>different climates</u>, and you'll be laughing if you get this stuff nailed.

Depressions and Anticyclones

Depressions are low pressure weather systems and anticyclones are high pressure weather systems —
they cause different weather. There's a depression or an anticyclone over the UK most of the time.

Depressions Form when Warm Air Meets Cold Air

Depressions form over the Atlantic ocean, then move east over the UK.
Here's how they form:

1) Warm, moist air from the tropics meets cold, dry air from the poles.

2) The warm air is less dense so it rises above the cold air.

3) Condensation occurs as the warm air rises,
   causing rain clouds to develop.

4) Rising air also causes low pressure at the Earth's surface.

5) So winds blow into the depression in a spiral (winds always
   blow from areas of high pressure to areas of low pressure).

6) A warm front is the front edge of the moving warm air.
   A cold front is the front edge of the moving cold air.

Depressions Cause a Sequence of Weather Conditions

You need to know the sequence of weather conditions that happen when a depression passes overhead.
Imagine you're stood on the ground ahead of the warm front and the depression's moving towards you.

	⑤ Cold air overhead	④ As the cold front passes	③ Warm air overhead	② As the warm front passes	① Ahead of the warm front
Rain	Showers	Heavy showers	None	Heavy	None
Clouds	High, broken	Towering, thick	None	Low, thick	High, thin
Pressure	Rising	Suddenly rising	Steady	Falling	Falling
Temperature	Cold	Falling	Warm	Rising	Cool
Wind speed	Decreasing	Strong	Decreasing	Strong	Increasing
Wind direction	NW	SW to NW	SW	SE to SW	SE

Wind direction is given as the direction the wind comes from.

Anticyclones cause Clear Skies and Dry Weather

Anticyclones also form over the Atlantic ocean and move east over the UK. Here's a bit about them:

1) Anticyclones are where air is falling, creating high pressure and light winds blowing outwards.

2) Falling air gets warmer so no clouds are formed, giving clear skies and no rain for days or even weeks.

3) In summer, anticyclones cause long periods of hot, dry, clear weather. There are no clouds to absorb
   the Sun's heat energy, so more gets through to the Earth's surface causing high temperatures.

4) In winter, anticyclones give long periods of cold, foggy weather. Heat is lost from the Earth's surface
   at night because there are no clouds to reflect it back. The temperature drops and condensation
   occurs near the surface, forming fog. (It doesn't heat up much in the day because the Sun is weak.)

I don't know about the weather, but I'm depressed after that page...

Depressions have a pretty appropriate name I think. They're complicated things too, so it's really worth taking your
time. Not only do you need to know the weather they cause, but also the reasons why (N.B. it's not just because they're mean).

Unit 1A — Weather and Climate

Extreme UK Weather

Extreme weather isn't as fun as extreme bog snorkelling, but it's becoming a lot more common in the UK.

Weather in the UK is Becoming More Extreme

1) It's raining more — the summer of 2007 was the wettest summer on record.
2) The rainfall is more intense, especially in winter — in some parts of Scotland the volume of rain that falls on wet winter days has gone up by 60%.
3) Temperature is increasing — the highest ever UK temperature was recorded in 2003 (38.5 °C).

More extreme weather has led to more extreme weather events in the last 10 years in the UK:

1) There was major flooding caused by storms and high rainfall in the south east in 2000, Cornwall in 2004, Cumbria in 2005, the Midlands in 2007 and Devon in 2008.
2) Strong winds (combined with high tides) caused flooding in Norfolk in 2006.
3) High temperatures led to a heatwave and drought conditions in the summer of 2003.

Extreme Weather has Impacts on lots of Different Things

©istockphoto.com /Andy Green

PEOPLE'S HOMES AND LIVES

* Floods damage homes and possessions, which can cost a lot to repair or replace.
* Businesses can be damaged by floods, so people can lose their income.
* Water use can be restricted during droughts, e.g. using hosepipes can be banned.
* Increased rainfall may mean water supplies are increased.

AGRICULTURE

* Droughts can cause crop failures.
* Increased rainfall can mean higher crop yields.
* A warmer climate means farmers can grow new crops, e.g. olives.

HEALTH

* Flooding can cause deaths by drowning.
* Heatwaves can cause deaths by heat exhaustion.
* Milder winters may reduce cold-related deaths.

TRANSPORT

* Floods can block roads and railways, disrupting transport systems.
* High temperatures can cause railway lines to buckle, so trains can't run properly.

Getting to work when it floods can be a mare.

There are Three ways of Reducing the Negative Impacts

1) PREPARING — individuals and local authorities can do things to prepare for extreme weather before it happens. For example, flood defences along rivers can be improved and education programmes can tell the public the best ways to cope with floods, droughts or heatwaves.
2) PLANNING — emergency services and local councils can plan how to deal with extreme weather events in advance, e.g. they can make plans for how to rescue people from floods and where to have shelters.
3) WARNING — warning systems give people time to prepare for extreme weather. For example, the Environment Agency issues flood warnings so people can prepare and evacuate.

UK weather — the latest thing on the extreme sports channel...

You need to know the impacts of extreme weather and the ways to reduce them. You also need to know some of the evidence for the weather becoming more extreme — and looking out the window at the dreary weather doesn't count.

Global Climate Change — Debate

We British like to talk about the weather, so global climate change should give us plenty to go on...

The Earth is Getting Warmer

Climate change is any change in the weather of an area over a long period. Global warming is the increase in global temperature over the last century. Global warming is a type of climate change and it causes other types of climate change, e.g. increased rainfall. Here's a bit about the evidence for global warming:

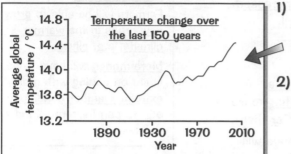

1) Global temperature has been measured using thermometers for the last 150 years. During the last 100 years, average global temperature has risen by about 0.9 °C. Average UK temperature has risen by about 1 °C.

2) Scientists have also reconstructed the climate over the last 1000 years using things like historical records, tree rings and cores taken from ice sheets.

3) This shows that global temperature is rising sharply now compared to how it was in the past.

4) There's some other evidence too:
   - The ice sheets are melting because global temperature is increasing — the Greenland Ice Sheet lost an average of 195 km³ of ice every year between 2003 and 2008.
   - Sea level is rising — increasing temperature causes ice on the land to melt and the oceans to expand. Sea level has risen 20 cm over the past century.

A few people argue that some of the evidence for global warming is a bit dodgy though. E.g. they say temperature measurements have shown an increase in temperature because human settlements have got closer to where a lot of the measurements are taken. Human settlements are warmer than natural environments because man-made surfaces like concrete absorb and radiate more heat energy.

An Increase in Greenhouse Gases is Causing Global Warming

There's a scientific consensus (general agreement) that global warming is caused by human activity:

1) An increase in human activities like burning fossil fuels, farming and deforestation has caused an increase in the concentration of carbon dioxide (CO_2) and methane (CH_4) in the atmosphere. For example, CO_2 has gone up from 280 ppm (parts per million) in 1850 to around 380 ppm today.

2) CO_2 and CH_4 are greenhouse gases — they trap heat reflected off the Earth's surface.

3) Greenhouse gases keep the Earth warm because they trap heat. Increasing the concentration of greenhouse gases in the atmosphere means the Earth heats up too much — causing global warming.

Here are some examples of other things that can cause climate change:

1) Variations in solar output — the Sun's output of energy isn't constant. In periods when there's more energy coming from the Sun, the Earth gets warmer.

2) Changes in the Earth's orbit — the way the Earth orbits the Sun changes, which affects how much energy the Earth receives. If the Earth receives more energy it gets warmer.

Which climate shall I change into today — the hot or the cold one...

The climate is changing — global warming is happening, it's just that a handful of people think some of the evidence isn't great. There are other things that cause climate change, but let's face it, we humans better take the rap this time.

Global Climate Change — Impacts

Global climate change will have <u>economic</u>, <u>social</u>, <u>environmental</u> and <u>political impacts</u> on the <u>world</u> and on the <u>UK</u>. Boy, that's <u>quite a few impacts</u>, but then, the climate's kind of important you know.

Climate Change will have Economic Impacts...

1) Climate change will affect <u>farming</u> in <u>different ways</u> around the <u>world</u>:
   - In <u>higher latitudes</u>, <u>warmer weather</u> will mean some farmers can make <u>more money</u> — some <u>crop yields</u> will be <u>increased</u>, and they'll be able to grow <u>new types</u> of crops to <u>sell</u>.
   - In <u>lower latitudes</u>, farmers' <u>income</u> may <u>decrease</u> because it's <u>too hot</u> and <u>dry</u> for farming.

2) Climate change means the <u>weather</u> is getting <u>more extreme</u>. This means <u>more money</u> will have to be <u>spent</u> on <u>predicting</u> extreme weather events, <u>reducing their impacts</u> and <u>rebuilding after</u> them.

3) <u>Industries</u> that help to <u>reduce the effects</u> of climate change will become <u>bigger</u> and <u>make more money</u>.

> **IN THE UK...**
> <u>Farmers</u> will be able to <u>grow new crops</u> in the <u>warmer climate</u>, e.g. <u>olives</u>.
> <u>More money</u> will have to be spent on coping with more <u>extreme weather conditions</u>, e.g. to pay for more flood defences.

...Social Impacts...

1) People <u>won't</u> be able to <u>grow as much food</u> in lower latitudes (see above). This could lead to <u>malnutrition</u>, <u>ill health</u> and <u>death</u> from <u>starvation</u>, e.g. in places like central Africa.

2) <u>More</u> people will <u>die</u> because of <u>more extreme weather events</u>.

3) <u>Hotter weather</u> makes it easier for some <u>infectious diseases</u> to <u>spread</u>. This will lead to more <u>ill health</u> and more <u>deaths from disease</u>.

4) Some areas will become so <u>hot</u> and <u>dry</u> that they're <u>uninhabitable</u>. People will have to <u>move</u>, which could lead to <u>overcrowding</u> in other areas.

> **IN THE UK...**
> There could be <u>fewer cold-related deaths</u>, but <u>more deaths</u> caused by <u>hot weather</u>, e.g. from <u>heat exhaustion</u>.
> <u>Diseases</u> that <u>don't exist</u> in the UK at the moment could <u>become common</u>, e.g. malaria.

Don't forget — global warming is a type of climate change.

...Environmental Impacts...

1) Global warming is causing <u>sea level</u> to <u>rise</u>, so some <u>habitats</u> will be <u>lost</u> as low-lying coastal environments are <u>submerged</u>.

2) <u>Rising temperature</u> and <u>decreased rainfall</u> will mean some environments will turn into <u>deserts</u>.

3) The <u>distribution</u> of some <u>species</u> may change due to climate change (species can only live in the areas where the <u>conditions suit them best</u>). Species that <u>can't move</u> may <u>die out</u>.

> **IN THE UK...**
> <u>Flooding</u> and <u>sea level rise</u> is <u>threatening</u> some <u>coastal habitats</u>, e.g. in the <u>south east</u> and <u>Norfolk</u>.
> The <u>distribution</u> of some <u>species</u> in the UK may change, e.g. it's thought <u>beech trees</u> will become <u>more common</u> in <u>Scotland</u>.

...and Political Impacts... phew

1) <u>Water</u> will become <u>more scarce</u> in <u>some places</u>. <u>Competition</u> over <u>water</u> could lead to <u>war</u> between countries.

2) Climate change may cause people to <u>move</u> (see above). This means some countries will have to cope with <u>increased immigration</u> and <u>emigration</u>.

3) Governments are <u>under pressure</u> to come up with <u>ways</u> to <u>slow climate change</u> or <u>reduce its effects</u>.

> **IN THE UK...**
> The government has had to set up a new <u>political department</u> to come up with ways to <u>slow climate change</u> and <u>reduce its impacts</u> — the <u>Department for Energy and Climate Change</u>.

<u>Olives in the garden and warmer weather — bring on global warming...</u>

The UK hasn't managed to escape <u>climate change</u>, so you need to make sure you learn the <u>impacts</u> on the <u>world</u> and the <u>UK too</u>. The impact on me right now is that I'm roasting hot and could do with a nice cool glass of fruit juice.

Global Climate Change — Responses

Most of the <u>responses</u> to climate change involve <u>cutting emissions</u> of <u>greenhouse gases</u> like CO_2. This can be done <u>globally</u>, <u>nationally</u> and <u>locally</u>, so everyone gets a slice of the fun.

The Kyoto Protocol is a Global Response

<u>Most countries</u> in the world have agreed to <u>monitor</u> and <u>cut greenhouse gas emissions</u> by signing an <u>international agreement</u> called the <u>Kyoto Protocol</u>:

1) The aim is to <u>reduce global</u> greenhouse gas emissions by <u>5% below 1990 levels</u> by <u>2012</u>.

2) Each country is set a <u>target</u>, e.g. the <u>UK</u> has agreed to reduce emissions by <u>12.5%</u> by 2012.

3) Another part of the protocol is the <u>carbon credits trading scheme</u>:

- <u>Countries</u> that come <u>under</u> their emissions target get <u>carbon credits</u> which they can <u>sell</u> to countries that <u>aren't meeting</u> their emissions target. This means there's a <u>reward</u> for having <u>low emissions</u>.

- <u>Countries</u> can also <u>earn</u> carbon credits by helping <u>poorer countries</u> to <u>reduce</u> their emissions. This means poorer countries will be able to reduce their emissions <u>more quickly</u>.

4) <u>Not all countries</u> have agreed to the Kyoto Protocol though — the <u>USA</u> hasn't agreed yet, and they have the <u>highest emissions</u> of any country in the world (<u>22%</u> of global CO_2 emissions in 2004).

There are also National and Local Responses to Climate Change

NATIONAL RESPONSES

1) **TRANSPORT STRATEGIES**
<u>Governments</u> can <u>improve public transport networks</u> like buses and trains. For example, they can make them run <u>faster</u> or <u>cover</u> a <u>wider area</u>. This <u>encourages more people</u> to <u>use public transport</u> instead of cars, so <u>CO_2</u> <u>emissions</u> are <u>reduced</u>.

2) **TAXATION**
Governments can <u>increase taxes</u> on cars with <u>high emissions</u>, e.g. in the UK there are <u>higher tax rates</u> for cars with <u>higher emissions</u>. This <u>encourages</u> people to <u>buy cars</u> with <u>low emissions</u>, so emissions are <u>reduced</u>.

LOCAL RESPONSES

1) **CONGESTION CHARGING**
<u>Local authorities</u> can <u>charge</u> people for <u>driving cars</u> into <u>cities</u> during <u>busy periods</u>, e.g. there's a congestion charge to drive into <u>central London</u> during busy times of the day. This <u>encourages</u> people to use their cars <u>less</u>, which <u>reduces emissions</u>.

2) **RECYCLING**
- Local authorities can <u>recycle more waste</u> by building <u>recycling plants</u> and giving people <u>recycling bins</u>. Recycling materials means <u>less energy</u> is used <u>making new materials</u>, so emissions are <u>reduced</u>.

- Local authorities can also <u>create energy</u> by <u>burning recycled waste</u>, e.g. Sheffield uses a waste incinerator to supply 140 buildings with energy.

3) **CONSERVING ENERGY**
- Local authorities give <u>money</u> and <u>advice</u> to make homes <u>more energy efficient</u>, e.g. by doing things like <u>improving insulation</u>. This means people <u>use less energy</u> to <u>heat</u> their homes, because <u>less</u> is <u>lost</u>. Emissions are <u>reduced</u> because <u>less energy</u> needs to be <u>produced</u>.

- <u>Individuals</u> can also conserve energy by doing things like <u>switching lights off</u> and <u>not</u> leaving electric gadgets on <u>standby</u>.

My response to climate change — slap on the suncream and bust out the shades...

Climate change is a <u>global problem</u>, so the response to deal with it needs to be on a <u>global scale</u>. That means <u>everyone</u> has to do their bit, from world leaders down to folk like us. Now, did I leave my hair straighteners on...

Tropical Storms

Tropical storms are <u>intense low pressure</u> weather systems. They've got lots of different names (<u>hurricanes</u>, <u>typhoons</u>, <u>tropical cyclones</u>, <u>tropical revolving storms</u> and <u>willy willies</u>), but they're all the <u>same thing</u>.

Tropical Storms Develop over Warm Water

Tropical storms are <u>huge storms</u> with <u>strong winds</u> and <u>torrential rain</u>. Scientists don't know exactly <u>how</u> they form, but they know <u>where</u> they form and some of the <u>conditions</u> that are <u>needed</u>:

1) Tropical storms develop above <u>sea water</u> that's <u>27 °C or higher</u>.

2) They happen in <u>late summer</u> and <u>autumn</u>, when sea temperatures are <u>highest</u>.

3) <u>Warm</u>, <u>moist</u> air <u>rises</u> and <u>condensation</u> occurs. This releases huge amounts of <u>energy</u>, which makes the storms <u>powerful</u>.

4) They <u>move west</u> because of the <u>easterly winds</u> near the equator.

5) They <u>lose strength</u> when they move over <u>land</u> because the energy supply from the warm water is <u>cut off</u>.

6) Most tropical storms occur between <u>5°</u> and <u>30°</u> north and south of the equator — any further from the equator and the water <u>isn't warm enough</u>.

7) The Earth's <u>rotation</u> deflects the path of the winds, which causes the storms to <u>spin</u>.

Tropic of Cancer (23° north)

Equator

Tropic of Capricorn (23° south)

→ path of tropical storm

sea surface temperature 27 °C or higher

You Need to Learn the Characteristics of Tropical Storms

Tropical storms <u>spin anticlockwise</u> and move <u>north west</u> (in the <u>northern hemisphere</u>). They're <u>circular</u> in shape, <u>hundreds of kilometres wide</u> and usually last between <u>7</u> and <u>14 days</u>. Here's a bit about their <u>structure</u> and the <u>weather conditions</u> they cause:

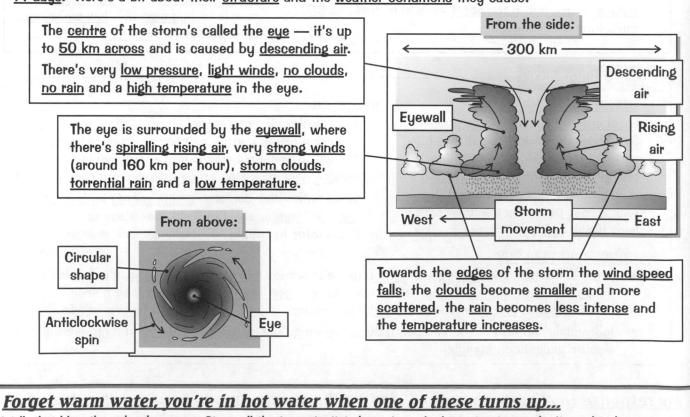

The <u>centre</u> of the storm's called the <u>eye</u> — it's up to <u>50 km across</u> and is caused by <u>descending air</u>. There's very <u>low pressure</u>, <u>light winds</u>, <u>no clouds</u>, <u>no rain</u> and a <u>high temperature</u> in the eye.

The eye is surrounded by the <u>eyewall</u>, where there's <u>spiralling rising air</u>, very <u>strong winds</u> (around 160 km per hour), <u>storm clouds</u>, <u>torrential rain</u> and a <u>low temperature</u>.

From the side:

← 300 km →

Descending air

Eyewall

Rising air

West ←

Storm movement

East

Towards the <u>edges</u> of the storm the <u>wind speed</u> <u>falls</u>, the <u>clouds</u> become <u>smaller</u> and more <u>scattered</u>, the <u>rain</u> becomes <u>less intense</u> and the <u>temperature increases</u>.

From above:

Circular shape

Anticlockwise spin

Eye

Forget warm water, you're in hot water when one of these turns up...

Well, that blew the cobwebs away. Since all the top scientists haven't worked it out yet, you don't need to know precisely how a tropical storm <u>forms</u>, you just need to know things that <u>cause</u> them and a bit about their <u>structure</u>.

Tropical Storms — Case Studies

Tropical storms can wreak quite a lot of havoc you know. Here are a couple of case studies...

Tropical Storms have Different Effects in Different Places

The effects of tropical storms and the responses to them are different in different parts of the world.
A lot depends on how wealthy the part of the world is. Here are a couple of case studies, so you can
cash in those marks when you get asked to compare two case studies in the exam.

Tropical storm in a
rich part of the world:

Name: Hurricane Katrina
Place: South east USA
Date: 29th August, 2005

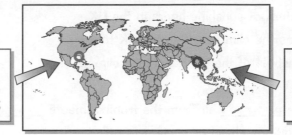

Tropical storm in a
poor part of the world:

Name: Cyclone Nargis
Place: Irrawaddy delta, Burma
Date: 2nd May, 2008

Preparation	• The USA has a sophisticated monitoring system to predict if a hurricane will hit (e.g. by using satellite images of the Atlantic) — so people were warned. • Mississippi and Louisiana declared states of emergency on 26th August — they set up control centres and stockpiled supplies. • 70-80% of New Orleans residents were evacuated before the hurricane reached land.	• Indian and Thai weather agencies warned the Burmese Government that Cyclone Nargis was likely to hit the country. Despite this, Burmese forecasters reported there was little or no risk. • There were no emergency or evacuation plans.
Social effects	• More than 1800 people were killed. • 300 000 houses were destroyed. • 3 million people left without electricity. • One of the main routes out of New Orleans was closed because parts of the I-10 bridge collapsed.	• More than 140 000 people were killed. • 450 000 houses were destroyed. • 2-3 million people were made homeless. • 1700 schools were destroyed.
Economic effects	• Total of around $300 billion of damage. • 230 000 of jobs were lost from businesses that were damaged or destroyed. • 30 offshore oil platforms sunk or went missing. This increased the price of fuel. • Shops in New Orleans were looted by residents in the days after the hurricane.	• Total of around $4 billion of damage. • Millions of people lost their livelihoods. • 200 000 farm animals were killed, crops were lost and over 40% of food stores were destroyed.
Environmental effects	• The hurricane caused the sea to flood parts of the land. This destroyed some coastal habitats, e.g. sea turtle breeding beaches.	• Coastal habitats such as mangrove forests were damaged. • The salinity (salt content) of soil in some areas has increased because of flooding by sea water. This means it's more difficult for plants to grow.
Short-term response	• During the storm the coast guard, police, fire service, army and volunteers rescued over 50 000 people. • About 25 000 people were given temporary shelter at a sports stadium (the Louisiana Superdome) immediately after the storm.	• Burma's Government initially refused to accept any foreign aid. Aid workers were only allowed in 3 weeks after the disaster occurred. • The UN launched a massive appeal to raise money to help respond to the disaster.
Long-term response	• The US government has spent over $800 million on rebuilding flood defences. • Around $34 billion has been set aside for the re-building of things like houses and schools.	• Burma is relying on international aid to repair the damage — fewer than 20 000 homes have been rebuilt and half a million survivors are still living in temporary shelters.

Katrina — sounds like a nice girl...

When you compare the two, it's clear that the impacts were worse in Burma and the long-term response has been a bit
more organised in the USA. Make sure you learn both case studies though — then you'll blow 'em away in the exam.

Unit 1A — Weather and Climate

Revision Summary for Weather and Climate

Well, wasn't that a blast of fresh air from the prevailing wind. Weather and climate is a pretty complicated topic, so don't worry if it didn't sink in first time round. Give these questions a bit of a whirl to see whether your weather knowledge is up to the task. If you're still in a hurricane of confusion, then look back through the topic and give the bits you don't know the once over — and don't let the depressions get you down.

1) During what months are temperatures highest in the UK?

2) Describe the seasonal pattern of sunshine hours in the UK.

3) Explain why latitude causes the climate within the UK to vary.

4) Give one reason why the whole of the UK has a mild climate.

5) Fill in the blanks below:

Depressions form when _____ _____ air from the tropics, meets _____ _____ air from the poles.

6) Describe the weather conditions ahead of a warm front.

7) Describe the weather conditions as a cold front passes overhead.

8) What weather conditions do anticyclones bring in winter?

9) Give one piece of evidence for the weather becoming more extreme in the UK.

10) Give two examples of extreme weather events that have happened in the UK in the last 10 years.

11) Give two impacts that extreme weather has on the homes and lives of people in the UK.

12) Describe one way that the negative impacts of extreme weather in the UK are being reduced.

13) What is global warming?

14) Give one piece of evidence for global warming.

15) How do greenhouse gases cause global warming?

16) Describe one cause of climate change (apart from greenhouse gases).

17) Give one economic impact of climate change on the world.

18) Give one social impact of climate change in the UK.

19) Give one environmental impact of climate change on the world.

20) Give one political impact of climate change in the UK.

21) How much does the Kyoto Protocol aim to reduce greenhouse gas emissions by?

22) Give one way that countries can earn carbon credits.

23) Describe one national response to climate change.

24) Describe one local response to climate change.

25) Give one condition that's needed for tropical storms to form.

26) Describe the distribution of tropical storms.

27) In what direction do tropical storms spin?

28) In what part of a tropical storm are there no clouds overhead?

29) a) Name one tropical storm that happened in a rich part of the world.

　　b) Give three effects of this tropical storm.

　　c) Give one long-term response to this tropical storm.

30) a) Name one tropical storm that happened in a poor part of the world.

　　b) Give three effects of this tropical storm.

　　c) Give one long-term response to this tropical storm.

Ecosystems

Welcome to a lovely new topic. Take your coat off, sit down and make yourself at home.
Do have a slice of cake — I made it myself. Just relax and I'll tell you all the gossip about <u>ecosystems</u>.

An Ecosystem Includes all the Living and Non-Living Parts in an Area

1) An <u>ecosystem</u> is a unit that includes all the <u>living parts</u> (e.g. plants and animals) and the <u>non-living (physical) parts</u> (e.g. soil and climate) in an <u>area</u>.

2) The <u>organisms</u> in ecosystems can be classed as <u>producers</u>, <u>consumers</u> or <u>decomposers</u>.

3) A <u>producer</u> is an organism that uses <u>sunlight energy</u> to <u>produce food</u>.

4) A <u>consumer</u> is an organism that gets its energy by <u>eating other organisms</u> — it eats <u>producers</u> or <u>other consumers</u>.

5) A <u>food chain</u> shows <u>what eats what</u>. A <u>food web</u> shows <u>lots of food chains</u> and how they <u>overlap</u>.

6) A <u>decomposer</u> is an organism that gets its energy by <u>breaking down dead material</u>, e.g. <u>dead producers</u>, <u>dead consumers</u> or <u>fallen leaves</u>. <u>Bacteria</u> and <u>fungi</u> are decomposers.

7) When <u>dead material</u> is <u>decomposed</u>, <u>nutrients</u> are <u>released</u> into the <u>soil</u>. The nutrients are then <u>taken up</u> from the soil <u>by plants</u>. The plants may be eaten by <u>consumers</u>. When the plants or consumers <u>die</u>, the <u>nutrients are returned</u> to the <u>soil</u>. This <u>transfer of nutrients</u> is called the <u>nutrient cycle</u>.

EXAMPLE of a SMALL SCALE ECOSYSTEM

- A <u>hedgerow</u> ecosystem includes the <u>plants</u> that make up the hedgerow, the <u>organisms that live in it</u> and <u>feed on it</u>, the <u>soil</u> in the area and the <u>rainfall</u> and <u>sunshine</u> it receives.

- The <u>producers</u> include <u>hawthorn bushes</u> and <u>blackberry bushes</u>.

- The <u>consumers</u> include <u>thrushes</u>, <u>ladybirds</u>, <u>spiders</u>, <u>greenfly</u>, <u>sparrows</u> and <u>sparrowhawks</u>.

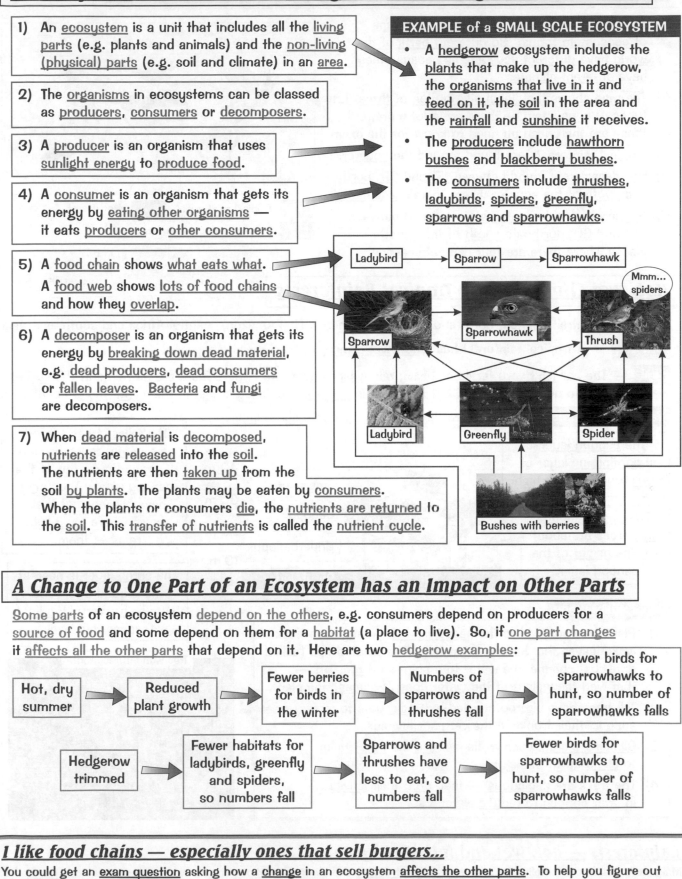

A Change to One Part of an Ecosystem has an Impact on Other Parts

<u>Some parts</u> of an ecosystem <u>depend on the others</u>, e.g. consumers depend on producers for a <u>source of food</u> and some depend on them for a <u>habitat</u> (a place to live). So, if <u>one part changes</u> it <u>affects all the other parts</u> that depend on it. Here are two <u>hedgerow examples</u>:

| Hot, dry summer | → | Reduced plant growth | → | Fewer berries for birds in the winter | → | Numbers of sparrows and thrushes fall | → | Fewer birds for sparrowhawks to hunt, so number of sparrowhawks falls |

| Hedgerow trimmed | → | Fewer habitats for ladybirds, greenfly and spiders, so numbers fall | → | Sparrows and thrushes have less to eat, so numbers fall | → | Fewer birds for sparrowhawks to hunt, so number of sparrowhawks falls |

I like food chains — especially ones that sell burgers...

You could get an <u>exam question</u> asking how a <u>change</u> in an ecosystem <u>affects the other parts</u>. To help you figure out the answer <u>draw</u> a <u>food web</u> for the ecosystem so you can easily see what will have <u>more food</u>, <u>less food</u>, <u>no habitat</u> etc.

World Ecosystems

I hope you've packed suitable clothes because I'm about to take you on a whistle-stop tour of the world's ecosystems. Actually, you don't need to know them all but you'll still need outfits for three of them...

You Need to Know the Global Distribution of Three Ecosystems

1) The climate in an area determines what type of ecosystem forms. So different parts of the world have different ecosystems because they have different climates.

2) The map shows the global distribution of three types of ecosystem — there are a lot more, but these are the ones you need to know for the exam.

3) Tropical rainforests are found around the equator.

4) Hot deserts are found between 15° and 30° north and south of the equator where there's less rainfall.

5) Temperate deciduous forests are found between 40° and 60° north and south of the equator in places where there are four distinct seasons.

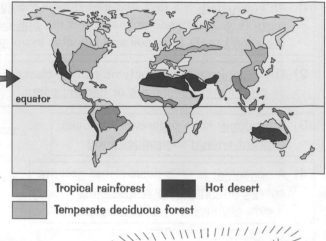

equator

| ■ Tropical rainforest | ■ Hot desert |
| ■ Temperate deciduous forest | |

See the next page for more on the other two types of ecosystem.

Hot, Wet Climates have Tropical Rainforests

Areas — Central America, north and east South America (the Amazon), central Africa and south east Asia.

Climate — A tropical rainforest has a hot, wet climate with no definite seasons.

Soil — The soil isn't very fertile as heavy rain washes nutrients away. There are nutrients at the surface due to decayed leaf fall, but this layer is very thin as decay is fast in the warm, moist conditions.

Vegetation structure — There are three tree layers and a shrub layer:

The canopy layer is a continuous layer of trees around 30 m high.

The undercanopy layer trees are about half the height of the canopy layer.

The tallest trees (called emergents) reach around 40 m and poke out of the canopy layer. They only have branches at their crown where most light reaches them.

The shrub layer is nearest the ground at around 10 m high. Very little light reaches this level.

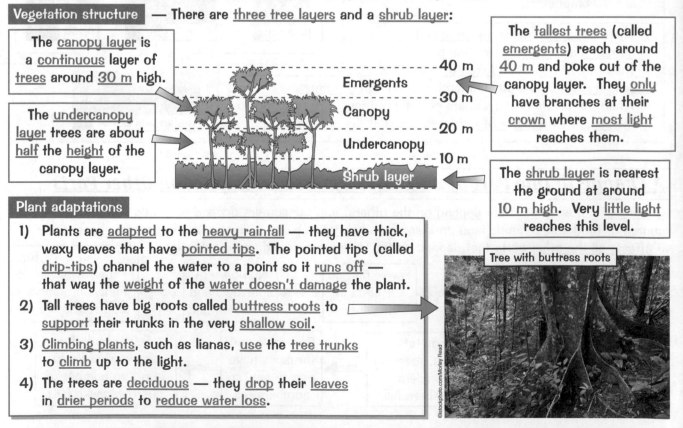

Emergents — 40 m
— 30 m
Canopy
— 20 m
Undercanopy
— 10 m
Shrub layer

Plant adaptations

1) Plants are adapted to the heavy rainfall — they have thick, waxy leaves that have pointed tips. The pointed tips (called drip-tips) channel the water to a point so it runs off — that way the weight of the water doesn't damage the plant.

2) Tall trees have big roots called buttress roots to support their trunks in the very shallow soil.

3) Climbing plants, such as lianas, use the tree trunks to climb up to the light.

4) The trees are deciduous — they drop their leaves in drier periods to reduce water loss.

Tree with buttress roots

©istockphoto.com/Morley Read

Rainforests — hot, wet and full of creepy bugs, eugh...

Make sure you know the climate, soil, vegetation structure and plant adaptations for the rainforest ecosystem. Cover the page and scribble down what you know to check. And give yourself bonus marks for getting the tree layer diagram right.

World Ecosystems

Bet you thought I'd forgotten about <u>hot deserts</u> and <u>temperate deciduous forests</u>... no such luck...

Hot, Dry Climates Have Hot Deserts

Areas — North Africa, the Middle East, south west USA, large parts of Australia.

Climate — There's very <u>little rainfall</u>. <u>When</u> it rains also <u>varies a lot</u> — it might only rain <u>once</u> every two or three years. <u>Temperatures</u> are <u>extreme</u> — they range from very <u>hot</u> in the <u>day</u> (e.g. 45 °C) to very <u>cold</u> at <u>night</u> (e.g. 5 °C).

Soil — It's usually <u>shallow</u> with a <u>coarse</u>, <u>gravelly texture</u>. There's <u>hardly any leaf fall</u> so the soil <u>isn't very fertile</u>.

Vegetation structure — Plant growth is pretty <u>sparse</u> due to a <u>lack of rainfall</u>. Plants that do grow include <u>cacti</u> and <u>thornbushes</u>.

Plant adaptations

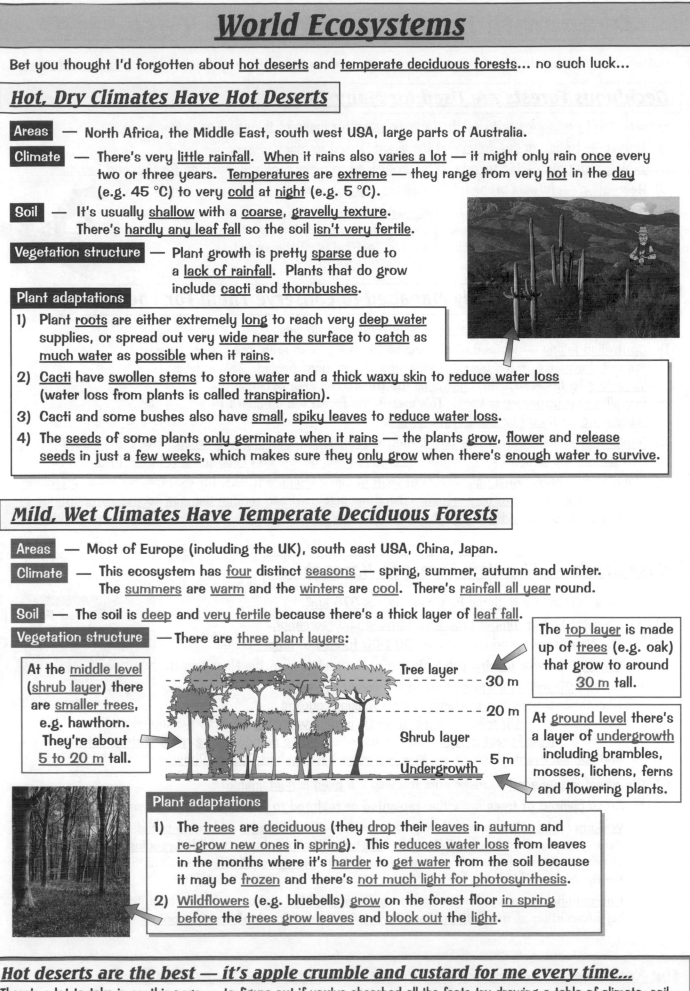

1) Plant <u>roots</u> are either extremely <u>long</u> to reach very <u>deep water</u> supplies, or spread out very <u>wide near the surface</u> to <u>catch</u> as <u>much water</u> as <u>possible</u> when it <u>rains</u>.

2) <u>Cacti</u> have <u>swollen stems</u> to <u>store water</u> and a <u>thick waxy skin</u> to <u>reduce water loss</u> (water loss from plants is called <u>transpiration</u>).

3) Cacti and some bushes also have <u>small</u>, <u>spiky leaves</u> to <u>reduce water loss</u>.

4) The <u>seeds</u> of some plants <u>only germinate when it rains</u> — the plants <u>grow</u>, <u>flower</u> and <u>release</u> <u>seeds</u> in just a <u>few weeks</u>, which makes sure they <u>only grow</u> when there's <u>enough water to survive</u>.

Mild, Wet Climates Have Temperate Deciduous Forests

Areas — Most of Europe (including the UK), south east USA, China, Japan.

Climate — This ecosystem has <u>four</u> distinct <u>seasons</u> — spring, summer, autumn and winter. The <u>summers</u> are <u>warm</u> and the <u>winters</u> are <u>cool</u>. There's <u>rainfall all year</u> round.

Soil — The soil is <u>deep</u> and <u>very fertile</u> because there's a thick layer of <u>leaf fall</u>.

Vegetation structure — There are <u>three plant layers</u>:

The <u>top layer</u> is made up of <u>trees</u> (e.g. oak) that grow to around <u>30 m</u> tall.

At the <u>middle level</u> (<u>shrub layer</u>) there are <u>smaller trees</u>, e.g. hawthorn. They're about <u>5 to 20 m</u> tall.

At <u>ground level</u> there's a layer of <u>undergrowth</u> including brambles, mosses, lichens, ferns and flowering plants.

Tree layer — 30 m
20 m
Shrub layer
Undergrowth — 5 m

Plant adaptations

1) The <u>trees</u> are <u>deciduous</u> (they <u>drop</u> their <u>leaves</u> in <u>autumn</u> and <u>re-grow new ones</u> in <u>spring</u>). This <u>reduces water loss</u> from leaves in the months where it's <u>harder</u> to <u>get water</u> from the soil because it may be <u>frozen</u> and there's <u>not much light for photosynthesis</u>.

2) <u>Wildflowers</u> (e.g. bluebells) <u>grow</u> on the forest floor <u>in spring</u> <u>before</u> the <u>trees grow leaves</u> and <u>block out</u> the <u>light</u>.

Hot deserts are the best — it's apple crumble and custard for me every time...

There's a lot to take in on this page — to figure out if you've absorbed all the facts try <u>drawing a table</u> of <u>climate</u>, <u>soil</u>, <u>vegetation structure</u> and <u>plant adaptations</u> for the <u>two ecosystems</u>. If there are any blank boxes read the page again.

Temperate Deciduous Forest — Case Study

Batten down the hatches, get your supplies ready and prepare for some details about <u>deciduous forests</u>...

Deciduous Forests are Used for Many Things

Forests don't just look pretty — they can be <u>used</u> for <u>loads of things</u>. For example:

1) <u>Timber</u> — <u>trees</u> are <u>cut down</u> and the <u>wood</u> is <u>sold</u> to <u>make money</u>.

2) <u>Timber products</u> — the <u>wood</u> can be <u>processed</u> to make <u>products</u> such as <u>fencing</u> and <u>furniture</u>.

3) <u>Recreation</u> — forests are used for <u>walking</u>, <u>cycling</u> and <u>other outdoor activities</u>.

If a forest is going to be <u>used</u> in the <u>long-term</u>, it has to be <u>managed</u> in a way that's <u>sustainable</u>, i.e. in a way that allows people <u>today</u> to get the things they need, but <u>without stopping</u> people in the <u>future</u> from getting what they <u>need</u>.

Forests can be Carefully Managed to Conserve Them For The Future

Here are some examples of <u>sustainable management strategies</u>:

1) <u>Controlled felling</u> — instead of clearing all the trees in an area, <u>only some trees</u> are <u>cut down</u>, e.g. trees <u>over</u> a <u>certain age</u> or just <u>one species</u>. This is <u>less damaging</u> to the forest than <u>felling all</u> the trees in an area because the <u>overall forest structure</u> is kept. This means the forest will be able to <u>regenerate</u> so it <u>can be used in the future</u>.

Controlled felling is also called selective logging.

2) <u>Replanting</u> — where trees are <u>felled</u>, they're <u>replaced</u> by <u>planting new trees</u>. This makes sure that <u>overall</u> the <u>amount</u> of <u>forest</u> is <u>not reduced</u> and people can <u>keep using it in the future</u>.

3) <u>Planning for recreational use</u> — lots of <u>visitors</u> can <u>damage</u> a forest, for example by causing <u>erosion</u>, <u>dropping litter</u> and <u>disturbing wildlife</u>. Good management can <u>reduce damage</u> so <u>use can continue in the future</u>, e.g. by <u>paving footpaths</u> to <u>reduce erosion</u> and <u>providing</u> plenty of <u>bins</u> to <u>reduce litter</u>.

Case Study — The New Forest in Hampshire

1) The New Forest is a <u>National Park</u> that covers <u>375 km²</u>.

2) It's used for <u>timber</u>, <u>timber products</u>, <u>farming</u> and <u>recreation</u>.

New Forest, Hampshire

- The New Forest produces around <u>50 000 tonnes</u> of <u>timber</u> a year.

- <u>Local mills</u> make <u>fencing products</u> out of the timber from the New Forest.

- Around <u>20 million visitors</u> come to the forest <u>per year</u>. Recreational <u>activities available</u> include <u>walking</u>, <u>cycling</u> (there are over <u>100 miles</u> of <u>cycle tracks</u>), <u>wildlife watching</u> (visitors particularly come to see the <u>New Forest ponies</u>, which roam wild), <u>horse riding</u>, <u>fishing</u>, <u>golf</u>, <u>watersports</u> and <u>special events</u> such as the <u>New Forest and Hampshire County Show</u>.

3) The forest is <u>managed</u> to make sure the way it's <u>used</u> is <u>sustainable</u>:

- Areas <u>cleared of trees</u> are either <u>replanted</u> or <u>restored to other habitats</u> like heathland.

- <u>Walkers</u> and <u>cyclists</u> are encouraged to stick to the <u>footpaths</u> and <u>cycle paths</u> to limit damage to surrounding <u>habitats</u>. Also <u>dogs aren't allowed near wildlife breeding sites</u> at certain times of year. These measures help to <u>conserve wildlife</u> so it's still there for future generations.

- <u>Recreational users</u> are encouraged to <u>act responsibly</u> (e.g. close gates, take litter home) by information at the <u>National Park Forest Centre</u> and <u>local information points</u>.

The New Forest — a bit like the old one but bigger, faster and shinier...

Eeee, the <u>New Forest sounds nice</u> — a magical forest with wild ponies and pigs roaming around. Try to keep that image in your head when you're scribbling out all the <u>details</u> about it in the <u>exam</u>. Which reminds me... <u>get learnin' the details</u>.

Tropical Rainforest — Deforestation

Removal of trees from forests is called deforestation. It's happening on a huge scale in many tropical rainforests. Deforestation has many impacts — some good, some bad and some downright ugly.

There are Five Main Causes of Deforestation

Farming — forest is cleared to set up small subsistence farms or large commercial cattle ranches. Often the "slash and burn" technique is used to clear the forest — vegetation is cut down and left to dry then burnt.

Mineral extraction — minerals (e.g. gold and iron ore) are mined and sold to make money. Trees are cut down to expose ground and to clear access routes.

Commercial logging — trees are felled to make money.

Population pressure — as the population in the area increases, trees are cleared to make land for new settlements.

Road building — more settlements and industry (e.g. logging and mining) lead to more roads being built. Trees along the path of the road have to be cleared to build them.

Deforestation has Environmental, Social, Political and Economic Impacts

ENVIRONMENTAL

1) Fewer trees means fewer habitats and food sources for animals and birds. This reduces biodiversity as organisms either have to move or become extinct.

2) With no trees to hold the soil together, heavy rain washes away the soil (soil erosion).

3) If a lot of soil from deforested areas is washed into rivers it can kill fish, make the water undrinkable and cause flooding (as the riverbed is raised so it can't hold as much water).

4) Without a tree canopy to intercept (catch) rainfall and tree roots to absorb it, more water reaches the soil. This increases the risk of flooding and reduces soil fertility as nutrients in the soil are washed down into the earth, out of reach of plants.

5) Without trees there's no leaf fall — so no nutrient supply to the soil, which makes it less fertile.

6) Trees remove CO_2 from the atmosphere when they photosynthesise, so without them less CO_2 is removed. Also, burning vegetation to clear forest produces CO_2. So deforestation means more CO_2 in the atmosphere, which adds to global warming.

7) Without trees, water isn't removed from the soil and evaporated into the atmosphere. So fewer clouds form and rainfall in the area is reduced. Reduced rainfall reduces plant growth.

SOCIAL

1) The quality of life for some local people improves as there are more jobs.

2) The livelihoods of some local people are destroyed — deforestation can cause the loss of the animals and plants that they rely on to make a living.

3) Some native tribes have been forced to move when trees on their land have been cleared.

4) There can be conflict between native people, landowners, mining companies and logging companies over use of land.

ECONOMIC

1) Logging, farming and mining create jobs.

2) A lot of money is made from selling timber, mining and commercial farming.

POLITICAL

There's pressure from foreign governments to stop deforestation.

The deforestation revision page — it's a cut above the rest...

A bit of a serious page, this one. I would try to brighten it up by doing a dance or something but you'd probably just point and laugh. So I'll stay sitting down and recommend you learn this page — deforestation is a fave with examiners.

Tropical Rainforest — Sustainable Management

It's <u>not</u> all <u>doom and gloom</u> for <u>rainforests</u>. In fact, this page is dedicated to the <u>ways to manage them</u>.

Tropical Rainforests can be Sustainably Managed

Rainforests can be managed in a way that's <u>sustainable</u>, i.e. in a way that allows people <u>today</u> to get the things they need, but <u>without stopping</u> people in the <u>future</u> from getting what they <u>need</u>. Here are some of the ways it can be done:

1) Selective Logging

1) Only <u>some trees</u> (e.g. just the <u>oldest ones</u>) are <u>felled</u> — <u>most trees</u> are <u>left standing</u>.

2) This is <u>less damaging</u> to the forest than <u>felling all</u> the trees in an area. If <u>only</u> a <u>few trees</u> are taken from each area the <u>overall forest structure</u> is <u>kept</u> — the canopy's still there and the soil isn't exposed. This means the forest <u>will</u> be able to <u>regenerate</u> so it can be <u>used in the future</u>.

3) The <u>least damaging forms</u> are 'horse logging' and 'helicopter logging' — <u>dragging</u> felled trees <u>out</u> of the <u>forest</u> using <u>horses</u> or removing them with <u>helicopters</u> instead of huge trucks.
EXAMPLE: <u>Helicopter logging</u> is used in the <u>Malaysian state</u> of <u>Sarawak</u>.

See the next page for more ways to manage rainforests sustainably.

3) Reducing Demand for Hardwood

1) <u>Hardwood</u> is a general term for wood from <u>certain tree species</u>, e.g. <u>mahogany</u> and <u>teak</u>. The wood tends to be <u>fairly dense</u> and <u>hard</u> — it's used to make things like <u>furniture</u>.

2) There's a <u>high demand</u> for hardwood from <u>consumers</u> in <u>richer countries</u>.

3) This means that <u>some tropical hardwood trees</u> are becoming <u>rarer</u> as people are chopping them down and selling them.

4) Some richer countries are trying to <u>reduce demand</u> so <u>fewer</u> of these tree species are <u>cut down</u>, which means they'll exist for <u>future generations to use</u>.

5) <u>Strategies</u> to reduce demand include <u>heavily taxing imported hardwood</u> or <u>banning its sale</u>.

6) Some countries with tropical rainforests also <u>ban logging</u> of <u>hardwood species</u>.

2) Replanting

1) This is when <u>new trees</u> are <u>planted</u> to <u>replace</u> the ones that are <u>cut down</u>.

2) This means there <u>will be trees</u> for people to <u>use</u> in the <u>future</u>.

3) It's important that the <u>same types of tree</u> are planted that were cut down, so that the <u>variety of trees</u> is <u>kept for the future</u>.

4) In some countries there are <u>environmental laws</u> to <u>make logging companies replant trees</u> when they clear an area.

4) Education

1) Some <u>local people don't know</u> what the <u>environmental impacts</u> of deforestation are. Local people try to make <u>money</u> in the <u>short-term</u> (e.g. by illegal logging) to <u>overcome</u> their own <u>poverty</u>.

2) Educating these people about the <u>impacts</u> of deforestation and <u>ways to reduce the impacts</u> <u>decreases their effect</u> on the <u>environment</u>.

3) Also, educating them about <u>alternative ways to make money</u> that <u>don't damage</u> the <u>environment</u>, e.g. ecotourism (see the next page), <u>reduces their impact</u>.

4) Both of these things mean that the rainforest is <u>conserved</u> and so will be there for <u>future generations</u> to <u>use</u>.

5) Education of the <u>international community</u> about the <u>impacts</u> of deforestation will <u>reduce demand</u> for <u>products</u> that <u>lead to deforestation</u>, e.g. hardwood furniture. It will also put <u>pressure</u> on <u>governments</u> to <u>reduce deforestation</u>.

Selective logging — a bit more precise than eeny, meeny, miny, mo...

Told you it wasn't all doom and gloom. You need to really <u>get your head around</u> what <u>sustainable management</u> is and how it makes sure that there are loads of trees, orangutans, tarantulas and other creepy bugs for future generations.

Tropical Rainforest — Sustainable Management

More rainforest <u>sustainable management strategies</u> for you to sink your teeth into...

Tropical Rainforests can be Sustainably Managed

1) Ecotourism

1) Ecotourism is <u>tourism</u> that <u>doesn't harm</u> the <u>environment</u> and <u>benefits</u> the <u>local people</u>.

2) Ecotourism provides a <u>source of income</u> for <u>local people</u>, e.g. they act as <u>guides</u>, <u>provide accommodation</u> and <u>transport</u>.

3) This means the local people <u>don't have to log</u> or <u>farm to make money</u>. So <u>fewer trees</u> are <u>cut down</u>, which means there are <u>more trees for the future</u>.

4) Ecotourism is usually a <u>small-scale</u> activity, with only <u>small numbers</u> of <u>visitors</u> going to an area at a time. This helps to keep the <u>environmental impact of tourism low</u>.

5) Ecotourism should cause as <u>little harm</u> to the <u>environment as possible</u>. For example, by making sure <u>waste</u> and <u>litter</u> are <u>disposed</u> of <u>properly</u> to prevent land and water <u>contamination</u>.

6) <u>Ecotourism helps</u> the <u>sustainable development</u> of an area because it <u>improves</u> the <u>quality of life</u> for <u>local people</u> <u>without stopping</u> people in the <u>future</u> getting what they <u>need</u> (because it doesn't damage the environment or deplete resources).

EXAMPLE: <u>Tataquara Lodge</u> is a tourist lodge in the <u>Brazilian rainforest</u>. The lodge has <u>15 rooms</u> and offers <u>activities</u> like <u>fishing</u>, <u>canoeing</u>, <u>wildlife viewing</u> and <u>forest walks</u>. <u>Waste</u> is <u>disposed of responsibly</u> and it <u>runs lights</u> using <u>solar power</u>.

See the previous page for a definition of sustainable management.

2) Reducing Debt

1) A lot of tropical rainforests are in <u>poorer countries</u>, e.g. Nigeria, Belize and Burma.

2) <u>Poorer countries</u> often <u>borrow money</u> from richer countries or organisations (e.g. the World Bank) to fund <u>development schemes</u> or <u>cope with emergencies</u> like floods.

3) This <u>money</u> has to be <u>paid back</u> (sometimes with <u>interest</u>).

4) These countries often <u>allow logging</u>, <u>farming</u> and <u>mining in rainforests</u> to <u>make money</u> to <u>pay back the debt</u>.

5) So <u>reducing debt</u> would mean countries <u>wouldn't have to do this</u> and the rainforests could be <u>conserved for the future</u>.

6) Debt can be <u>cancelled</u> by countries or organisations, but there's <u>no guarantee</u> the <u>money will be spent on conservation</u>.

7) <u>Conservation swaps</u> (debt-for-nature swaps) guarantee the <u>money is spent on conservation</u> — part of a country's debt is <u>paid off</u> by someone else in <u>exchange</u> for <u>investment</u> in <u>conservation</u>.

EXAMPLE: In <u>1987</u> a <u>conservation group</u> paid off some of <u>Bolivia's debt</u> in exchange for <u>creating a rainforest reserve</u>.

3) Protection

1) <u>Environmental laws</u> can be used to <u>protect rainforests</u>. For example:
   - Laws that <u>ban</u> the use of wood from forests that are managed <u>non-sustainably</u>.
   - Laws that <u>ban illegal logging</u>.
   - Laws that <u>ban logging</u> of <u>some tree species</u>, e.g. mahogany.

2) Many <u>countries</u> have set up <u>national parks</u> and <u>nature reserves</u> within rainforests. In these areas <u>damaging activities</u>, e.g. logging, are <u>restricted</u>. However, a <u>lack</u> of <u>funds</u> can make it <u>difficult</u> to <u>police</u> the restrictions.

Rainforest protection — you'll need an umbrella and some wellies...

Don't forget — the basic idea is that <u>anything</u> that allows <u>people today</u> to get what they <u>need</u> whilst <u>stopping</u> the <u>rainforest being damaged</u> or its resources being <u>depleted</u> is <u>sustainable management</u>.

Tropical Rainforest — Case Study

The Amazon is the largest rainforest on Earth, but it's shrinking fast due to deforestation.

Deforestation is a Problem in the Amazon

The Amazon covers an area of around 8 million km², including parts of Brazil, Peru, Colombia, Venezuela, Ecuador, Bolivia, Guyana, Suriname and French Guiana. However, since 1970 over 600 000 km² has been destroyed by deforestation. There are lots of causes — for example, between 2000 and 2005:

1) 60% was caused by cattle ranching.

2) 33% was caused by small-scale subsistence farming.

3) 3% was caused by logging.

4) 3% was caused by mining, urbanisation, road construction, dams and fires.

5) 1% was caused by large-scale commercial farming (other than cattle ranching).

Amazon Rainforest

South America

Deforestation in the Amazon has Many Impacts

Environmental
- Habitat destruction and loss of biodiversity, e.g. the number of endangered species in Brazil increased from 218 in 1989 to 628 in 2008.
- The Amazon stores around 100 billion tonnes of carbon — deforestation will release some of this as carbon dioxide, which causes global warming.

Social
- Local ways of life have been affected, e.g. some Brazilian rubber tappers have lost their livelihoods as rubber trees have been cut down.
- Native tribes have been forced to move, e.g. some of the Guarani tribe in Brazil have moved because their land was taken for cattle ranching and sugar plantations.
- There's conflict between large landowners, subsistence farmers and native people, e.g. in 2009 there were riots in Peru over rainforest destruction and hundreds of native Indians were killed or injured.

Economic
- Farming makes a lot of money for countries in the rainforest, e.g. in 2008, Brazil made $6.9 billion from trading cattle.
- The mining industry creates jobs for loads of people, e.g. the Buenaventura Mining Company in Peru employs over 3100 people.

Several Sustainable Management Strategies are being Used

1) Some deforested areas are being replanted with new trees, e.g. Peru plans to replant more than 100 000 km² of forest before 2018.

2) Some countries are trying to reduce the number of hardwood trees felled, e.g. Brazil banned mahogany logging in 2001 and seizes timber from illegal logging companies.

3) Ecotourism is becoming more popular, e.g. the Madre de Dios region in Peru has around 70 lodges for ecotourists — 60 000 people visited the region in 2007.

4) Most countries have environmental laws to help protect the rainforest, e.g. the Brazilian Forest Code says that landowners have to keep 50-80% of their land as forest.

5) Some countries have national parks, e.g. the Central Amazon Conservation Complex in Brazil is the largest protected area in the rainforest, covering around 25 000 km². It's a World Heritage Site that's home to loads of ecosystems and animals like black caimans and river dolphins.

6) Reducing debt has helped some countries conserve their rainforest, e.g. in 2008 the USA reduced Peru's debt by $25 million in exchange for conserving its rainforest.

Tarzan wouldn't have much fun in the Amazon...

At this point I'm supposed to write something helpful, but it turns out I'm a bit short of ideas. I've even tried making a cup of tea to help the ideas flow. I'll be prepared with a nugget of wisdom by the time you finish page 39 though.

Hot Deserts — Case Study

Hot deserts aren't just places for <u>riding camels</u> — believe it or not, they're <u>used for loads of stuff</u>.

Hot Deserts Provide Economic Opportunities

1) <u>Hot deserts</u> exist in <u>rich</u> and <u>poor areas</u> of the world.

2) Hot deserts in <u>rich areas</u> are usually used for things like <u>commercial farming</u>, <u>mining</u> and <u>tourism</u>. Lots of people also <u>retire there</u> (<u>retirement migration</u>).

3) Hot deserts in <u>poor areas</u> are usually used for <u>hunting and gathering</u> and <u>farming</u>.

4) <u>Management</u> of both rich and poor deserts needs to be <u>sustainable</u> — i.e. to allow people <u>today</u> to get the things they need, but <u>without stopping</u> people in the <u>future</u> from getting what they <u>need</u>.

See the next page for another hot desert case study.

Case Study — The Kalahari Desert is a Relatively Poor Region

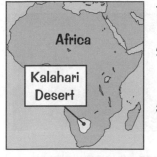

1) The <u>Kalahari Desert</u> has an area of <u>260 000 km²</u>. It covers most of <u>Botswana</u> and parts of <u>Zimbabwe</u>, <u>Namibia</u> and <u>South Africa</u>.

2) It gets <u>little rain</u> (about <u>200 mm per year</u>). The only <u>permanent river</u> in the area is the <u>Boteti River</u>. However, <u>temporary streams</u> and <u>rivers</u> form <u>after rain</u>. The <u>low rainfall</u> in the area means that <u>droughts</u> are a <u>problem</u>.

3) The Kalahari is very <u>sparsely populated</u>, but there are <u>native people</u> that live there, e.g. the <u>San Bushmen</u> and the <u>Tswana</u>. Some native people still <u>hunt wild game</u> (e.g. antelope) with bows and arrows and <u>gather plants</u> for <u>food</u>.

4) <u>Farming</u> cattle, goats and sheep is a <u>big industry</u> in the Kalahari, e.g. in <u>1998</u> there were <u>2.3 million cattle</u> in <u>Botswana</u>. Some grazing land is <u>irrigated</u> using <u>groundwater</u> from <u>boreholes</u>.

5) There's lots of <u>mining</u> in the area — there are <u>coal</u>, <u>diamond</u>, <u>gold</u>, <u>copper</u>, <u>nickel</u> and <u>uranium mines</u>, e.g. the <u>Opara Diamond Mine</u> in <u>Botswana</u>.

6) Some <u>uses</u> of the Kalahari have <u>negative impacts</u>:

> 1) <u>Overgrazing</u> of land has caused <u>soil erosion</u>, and <u>irrigation</u> has <u>depleted groundwater supplies</u>.
>
> 2) <u>Fences</u> put up by farmers have <u>blocked migration routes</u> of <u>wild animals</u>, e.g. <u>wildebeest</u>. The animals can't move to where the grazing is best so some die from starvation.
>
> 3) <u>Mining</u> and <u>farming</u> have led to <u>native people</u> being <u>forced off their land</u>.
>
> 4) <u>Mining uses a lot of water</u> from boreholes. This is <u>depleting groundwater supplies</u>.

7) Here are a few of the <u>management strategies</u> being carried out in the Kalahari:

> 1) Some places are trying to <u>conserve water</u>. E.g. in <u>Windhoek</u> in <u>Namibia</u> people are <u>charged</u> for the <u>volume of water they use</u>. This <u>encourages them to use less</u>. This is <u>more sustainable</u> because <u>water supplies aren't depleted as much</u> and so more will be there in the future.
>
> 2) <u>Water supply</u> all over the Kalahari is being increased by <u>building dams</u> and <u>drilling more boreholes</u>. This allows <u>more farming</u> and <u>reduces</u> the <u>effects of drought</u>, but <u>isn't sustainable</u> because it <u>depletes groundwater supplies</u> even more.
>
> 3) Several <u>game reserves</u> have been created to <u>provide areas</u> for the <u>native people to live</u> and to <u>protect wildlife</u>. For example, the <u>Central Kalahari Game Reserve</u> in <u>Botswana</u> was set up in 1961 as a <u>refuge</u> for the <u>San Bushmen</u>. This <u>is sustainable</u> because it <u>conserves</u> the way of life of the <u>native people</u> and <u>conserves the wildlife</u> for <u>future generations</u>.
>
> 4) Some agricultural <u>fences</u> have <u>been removed</u> to <u>allow animals</u> to <u>migrate</u>. This <u>is sustainable</u> because <u>fewer wild animals die</u> so they will still be around in the future.

The Kalahari Desert — isn't that full of squid...

I know, I know, <u>another case study</u>. They're absolutely everywhere I'm afraid. That doesn't mean you should skip over learning them though. <u>Cover the page</u> and <u>see what's stuck</u> in your head. Better to find out what you know now...

Hot Deserts — Case Study

Keep going my desert friend, just <u>one more case study</u> to go before the oasis.

Case Study — The Mojave Desert is a Relatively Rich Region

1) The <u>Mojave Desert</u> in the <u>USA</u> covers over <u>57 000 km²</u> and includes parts of <u>California</u>, <u>Nevada</u>, <u>Utah</u> and <u>Arizona</u>.

2) The region gets <u>less than 250 mm</u> of <u>rain</u> per <u>year</u>.

3) There's <u>commercial farming</u> in the area. For example, there have been <u>cattle ranches</u> in the region for <u>over 100 years</u>.

4) The area is <u>sparsely populated</u> but the <u>population</u> is <u>increasing</u>, e.g. <u>Las Vegas</u> in Nevada is the USA's <u>fastest growing city</u>. The area is popular with <u>people retiring</u> due to its year-round <u>good weather</u>, e.g. <u>80%</u> of the people in <u>Sun City</u> in Arizona are <u>over 65</u>.

5) <u>Water</u> for farming and for people comes from <u>groundwater</u>, the <u>Mojave River</u> and the <u>Colorado River</u>.

6) The region has many <u>tourist destinations</u> including <u>Las Vegas</u>, <u>Death Valley</u> and the <u>Grand Canyon</u>. The <u>Death Valley National Park</u> gets around <u>1 million visitors</u> per year. Tourists are attracted by the <u>wildlife</u> and <u>geology</u>, and activities like <u>camping</u>, <u>hiking</u>, <u>horse riding</u> and <u>off-road driving</u>.

7) In the past, <u>gold</u>, <u>silver</u>, <u>copper</u>, <u>lead</u> and <u>salts</u> were <u>mined</u>, although most mines have now <u>closed</u>. There are a few <u>borax</u> mines <u>still working</u> in <u>California</u>.

8) Some <u>uses</u> of the Mojave have <u>negative impacts</u>:

> 1) <u>Rapid population growth</u> (including retirement migrants) has <u>depleted water resources</u>.
>
> 2) <u>Farming uses a lot of water</u>, and it can also cause <u>soil erosion</u>.
>
> 3) <u>Tourists deplete water resources</u>, <u>drop litter</u>, <u>damage plants</u> and cause <u>soil erosion</u> (e.g. by using <u>off-road vehicles</u>).

9) Here are a few of the <u>management strategies</u> being carried out in the Mojave:

> 1) There are <u>water conservation schemes</u> in the area, e.g. the <u>Mojave Water Agency</u> gives people <u>vouchers</u> to buy <u>water efficient toilets</u> and <u>washing machines</u>. They also <u>pay people</u> to <u>remove grass lawns</u> (which need a lot of water) and replace them with plants that don't use as much water. These things are <u>more sustainable</u> because they <u>don't deplete water supplies as much</u>, so there's <u>more</u> for <u>future generations</u>.
>
> 2) The Mojave Desert has <u>four National Parks</u> (<u>Death Valley</u>, <u>Joshua Tree</u>, <u>Zion</u> and the <u>Grand Canyon</u>). <u>Native species</u> are <u>protected</u> and there are <u>strict rules</u> on <u>land use</u>, e.g. there are strict rules on <u>mining</u> to <u>reduce environmental damage</u>. This <u>is sustainable</u> because it <u>conserves the area</u>, so <u>future generations can use it</u>.
>
> 3) There are <u>designated roads</u> for <u>off-road vehicles</u>, and <u>sensitive areas</u> are <u>fenced off</u> so they <u>can't get in</u>. This <u>is sustainable</u> because it helps <u>conserve the plant life</u> for future generations.
>
> 4) Some <u>hotels</u> in <u>Las Vegas</u> are trying to <u>conserve water</u>, e.g. the <u>MGM Mirage®</u> Hotels use <u>drip-irrigation</u> to water lawns. This is <u>more sustainable</u> as it <u>doesn't use as much water</u> as other irrigation methods, so conserves more water for the future.

Death Valley — sounds like a laugh a minute...

Mojave (said 'mo-har-ve') means 'the meadows', which is a bit weird as not much grows there. It's also a bit weird that something like <u>water</u> can <u>run out</u>, but <u>hot deserts don't get a lot</u> of it so if people use it up, it might not get replaced.

Revision Summary for The Living World

So now you know absolutely everything there is to know about ecosystems. Or at least you know everything you need to for the exam. But before you go rushing off to celebrate with a slice of pumpkin pie and an Irish jig, best make sure you actually do know it. Now's as good a time as any, so give these questions a go.

1) Define the term ecosystem.

2) What is a consumer?

3) What is a food web?

4) Describe how nutrients are transferred to the soil in an ecosystem.

5) Describe the global distribution of tropical rainforests.

6) How many layers of vegetation does a tropical rainforest have?

7) What is an emergent tree?

8) Give three ways rainforest plants are adapted to their environment.

9) Describe the soil in a hot desert.

10) Give two ways plants are adapted to the hot desert environment.

11) Describe the climate of a temperate deciduous forest.

12) How tall is the top tree layer in a temperate deciduous forest?

13) What is controlled felling?

14) a) Give an example of a temperate deciduous forest.

    b) Describe how the forest is used for recreation.

    c) Describe how the forest is managed to make sure the way it's used is sustainable.

15) What are the five main causes of rainforest deforestation?

16) Give three environmental impacts of rainforest deforestation.

17) Give two social impacts of rainforest deforestation.

18) What is selective logging?

19) What is replanting?

20) How does reducing demand for hardwood help to conserve rainforests?

21) How can education be used to reduce rainforest deforestation?

22) What is ecotourism?

23) Give one way a country can reduce its debt in order to reduce deforestation.

24) Give two ways a country can protect its rainforest.

25) a) Give an example of a tropical rainforest.

    b) Describe one social, one economic and one environmental impact of deforestation in that rainforest.

    c) Give three ways the rainforest is being sustainably managed.

26) a) Give an example of a hot desert in a rich part of the world and a hot desert in a poorer part of the world.

    b) Compare the way the two desert areas are used.

    c) How is the desert in the rich area sustainably managed?

Unit 1B — Water on the Land

The River Valley

You're probably best off going to the loo before you start this topic. It's all about <u>flowing water</u>...

A River's Long Profile and Cross Profile Vary Over its Course

1) The <u>path</u> of a river as it <u>flows downhill</u> is called its <u>course</u>.

2) Rivers have an <u>upper course</u> (closest to the <u>source</u> of the river), a <u>middle course</u> and a <u>lower course</u> (closest to the <u>mouth</u> of the river).

3) Rivers form <u>channels</u> and <u>valleys</u> as they <u>flow downhill</u>.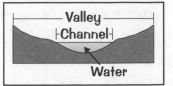

4) They <u>erode</u> the landscape — <u>wear it down</u>, then <u>transport</u> the material to somewhere else where it's <u>deposited</u>.

5) The <u>shape</u> of the <u>valley</u> and <u>channel changes</u> along the river depending on whether <u>erosion</u> or <u>deposition</u> is having the <u>most impact</u> (is the <u>dominant process</u>).

6) The <u>long profile</u> of a river shows you how the <u>gradient</u> (steepness) <u>changes</u> over the different courses.

7) The <u>cross profile</u> shows you what a <u>cross-section</u> of the river looks like.

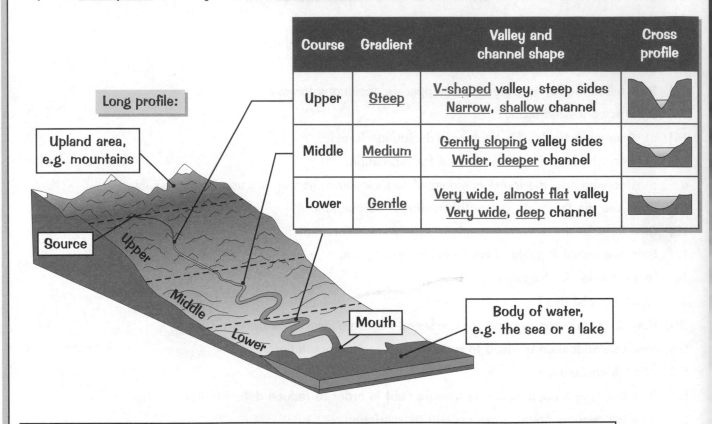

Course	Gradient	Valley and channel shape	Cross profile
Upper	Steep	<u>V-shaped</u> valley, steep sides <u>Narrow</u>, <u>shallow</u> channel	
Middle	Medium	<u>Gently sloping</u> valley sides <u>Wider</u>, <u>deeper</u> channel	
Lower	Gentle	<u>Very wide</u>, almost <u>flat</u> valley <u>Very wide</u>, <u>deep</u> channel	

Vertical and Lateral Erosion Change the Cross Profile of a River

Erosion can be <u>vertical</u> or <u>lateral</u> — both types happen at the <u>same time</u>, but one is usually <u>dominant</u> over the other at <u>different points</u> along the river:

Vertical erosion

This <u>deepens</u> the river valley (and channel), making it <u>V-shaped</u>. It's dominant in the <u>upper course</u> of the river.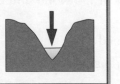

Lateral erosion

This <u>widens</u> the river valley (and channel). It's dominant in the <u>middle</u> and <u>lower courses</u>.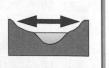

The river valet was rubbish at his job — all the cars got soaked...

Don't stress when revising this stuff — rivers are peaceful and calming for the soul. Hmm... just make sure you can describe a river's <u>long profile</u> and its <u>cross profile</u> at different courses, then you'll be calm and peaceful after the exam.

Erosion, Transportation and Deposition

Rivers <u>scrape</u> and <u>smash rocks up</u>, <u>push</u> them about, then <u>dump them</u> when they've had enough...

There are Four Processes of Erosion

1) <u>Hydraulic action</u>

The <u>force</u> of the water <u>breaks rock particles away</u> from the <u>river channel</u>.

I'm taking hydraulic action against this revision.

2) <u>Abrasion</u>

Eroded <u>rocks</u> picked up by the river <u>scrape</u> and <u>rub</u> against the <u>channel</u>, wearing it away. <u>Most erosion</u> happens by <u>abrasion</u>.

3) <u>Attrition</u>

Eroded <u>rocks</u> picked up by the river <u>smash into each other</u> and break into <u>smaller fragments</u>. Their <u>edges</u> also get <u>rounded off</u> as they rub together.

4) <u>Solution</u>

River water <u>dissolves</u> some types of rock, e.g. <u>chalk</u> and <u>limestone</u>.

The faster a river's flowing, the more erosion happens.

Transportation is the Movement of Eroded Material

The <u>material</u> a river has <u>eroded</u> is <u>transported downstream</u>.
There are <u>four processes</u> of transportation:

1) <u>Traction</u> — <u>large</u> particles like boulders are <u>pushed</u> along the <u>river bed</u> by the <u>force of the water</u>.

2) <u>Saltation</u> — <u>pebble-sized</u> particles are <u>bounced along</u> the <u>river bed</u> by the <u>force of the water</u>.

3) <u>Suspension</u> — <u>small</u> particles like silt and clay are <u>carried along</u> by the water.

4) <u>Solution</u> — <u>soluble materials</u> <u>dissolve</u> in the water and are <u>carried along</u>.

Traction · Saltation · Suspension · Solution

Deposition is When a River Drops Eroded Material

1) Deposition is when a river <u>drops</u> the <u>eroded material</u> it's <u>transporting</u>.
2) It happens when a river <u>slows down</u> (<u>loses velocity</u>).
3) There are a <u>few reasons</u> why rivers slow down and deposit material:

- The <u>volume</u> of <u>water</u> in the river <u>falls</u>.
- The <u>amount</u> of <u>eroded material</u> in the water <u>increases</u>.
- The water is <u>shallower</u>, e.g. on the <u>inside of a bend</u>.
- The river <u>reaches</u> its <u>mouth</u>.

In rock school there's only one punishment for naughty clay — suspension...

There are loads of amazingly similar names to remember here — try not to confuse <u>saltation</u>, <u>solution</u> and <u>suspension</u>. And yes, <u>solution</u> is both a process of <u>erosion</u> and <u>transportation</u>. Now saltate on over to the next page...

Unit 1B — Water on the Land

River Landforms

When a river's <u>eroding</u> and <u>depositing</u> material, <u>meanders</u> and <u>ox-bow lakes</u> can form. Australians have a different name for <u>ox-bow lakes</u> — billabongs. Stay tuned for more incredible facts.

Meanders are Formed by Erosion and Deposition

In their <u>middle</u> and <u>lower courses</u>, rivers develop <u>large bends</u> called <u>meanders</u>:

1) The <u>current</u> (the flow of the water) is <u>faster</u> on the <u>outside</u> of the bend because the river channel is <u>deeper</u> (there's <u>less friction</u> to <u>slow</u> the water down).

2) So more <u>erosion</u> takes place on the <u>outside</u> of the bend, forming <u>river cliffs</u>.

3) The <u>current</u> is <u>slower</u> on the <u>inside</u> of the bend because the river channel is <u>shallower</u> (there's <u>more friction</u> to <u>slow</u> the water down).

4) So eroded material is <u>deposited</u> on the <u>inside</u> of the bend, forming <u>slip-off slopes</u>.

The Mississippi River in the USA has lots of meanders.

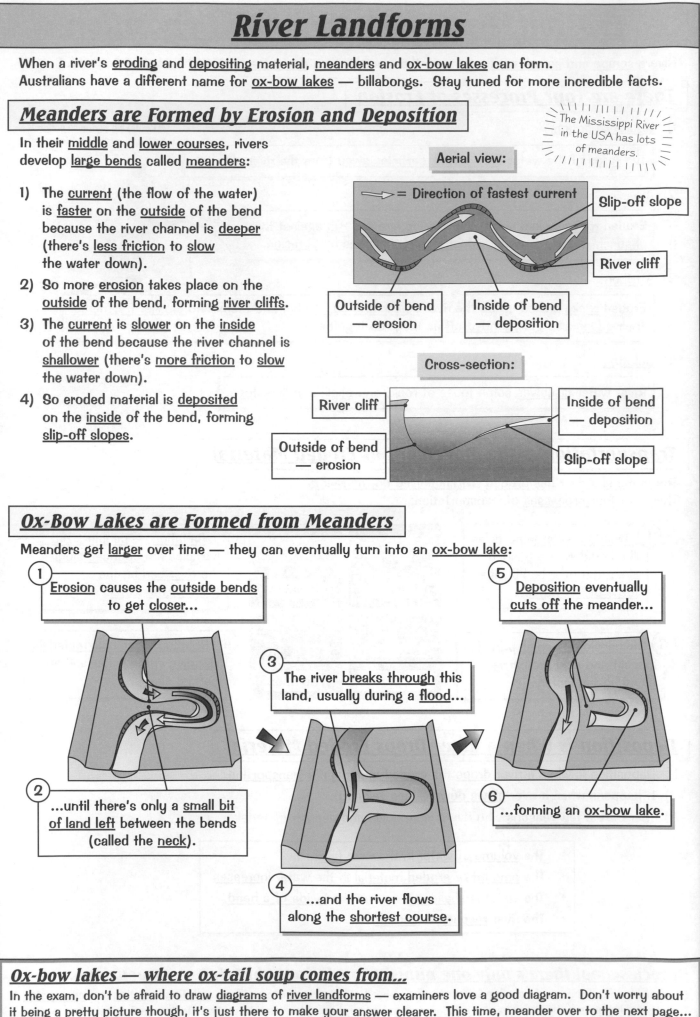

Aerial view:

⇒ = Direction of fastest current

Slip-off slope

River cliff

Outside of bend — erosion

Inside of bend — deposition

Cross-section:

River cliff

Outside of bend — erosion

Inside of bend — deposition

Slip-off slope

Ox-Bow Lakes are Formed from Meanders

Meanders get <u>larger</u> over time — they can eventually turn into an <u>ox-bow lake</u>:

① <u>Erosion</u> causes the <u>outside bends</u> to get <u>closer</u>...

② ...until there's only a <u>small bit of land left</u> between the bends (called the <u>neck</u>).

③ The river <u>breaks through</u> this land, usually during a <u>flood</u>...

④ ...and the river flows along the <u>shortest course</u>.

⑤ <u>Deposition</u> eventually <u>cuts off</u> the meander...

⑥ ...forming an <u>ox-bow lake</u>.

Ox-bow lakes — where ox-tail soup comes from...

In the exam, don't be afraid to draw <u>diagrams</u> of <u>river landforms</u> — examiners love a good diagram. Don't worry about it being a pretty picture though, it's just there to make your answer clearer. This time, meander over to the next page...

River Landforms

Four more <u>landforms</u> to sink your teeth into — mmm...

Waterfalls and Gorges are Formed by Erosion

1) <u>Waterfalls</u> (e.g. High Force waterfall on the River Tees) form where a river flows over an area of <u>hard rock</u> followed by an area of <u>softer rock</u>.

2) The <u>softer rock</u> is <u>eroded more</u> than the <u>hard rock</u>, creating a '<u>step</u>' in the river.

3) As water goes over the step it <u>erodes more and more</u> of the softer rock.

4) A <u>steep drop</u> is eventually created, which is called a <u>waterfall</u>.

5) The <u>hard rock</u> is eventually <u>undercut</u> by erosion. It becomes <u>unsupported</u> and <u>collapses</u>.

6) The collapsed rocks are <u>swirled around</u> at the foot of the waterfall where they <u>erode</u> the softer rock by <u>abrasion</u> (see p. 43). This creates a deep <u>plunge pool</u>.

7) Over time, <u>more undercutting</u> causes <u>more collapses</u>. The waterfall will <u>retreat</u> (move back up the channel), leaving behind a steep-sided <u>gorge</u>.

Waterfalls are found in the upper course of a river.

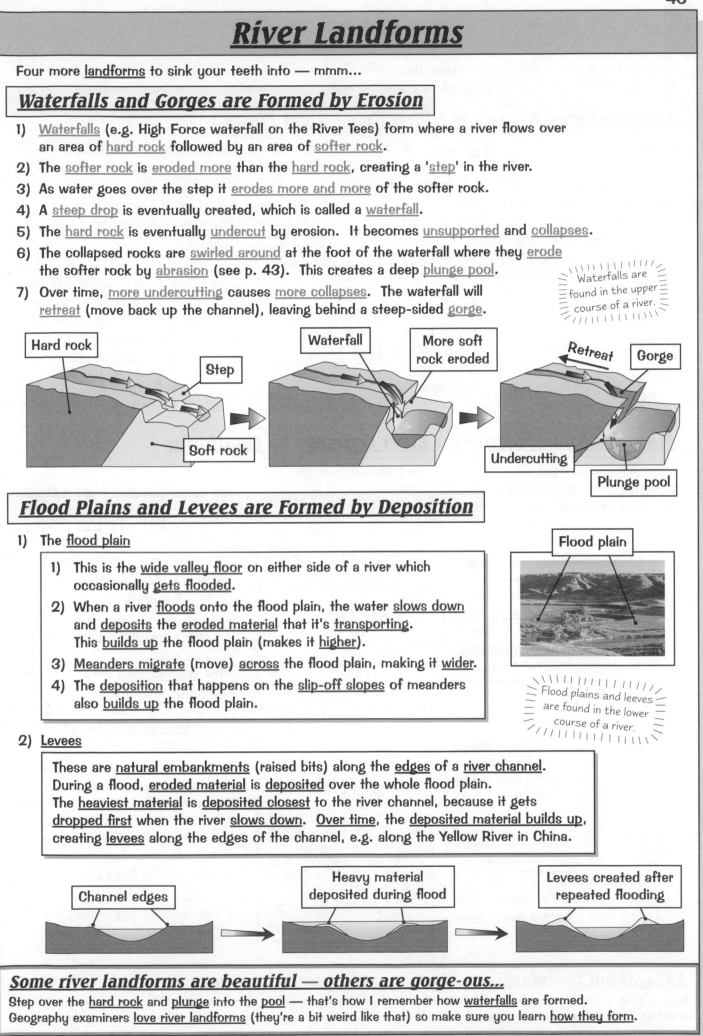

Flood Plains and Levees are Formed by Deposition

1) The <u>flood plain</u>

> 1) This is the <u>wide valley floor</u> on either side of a river which occasionally <u>gets flooded</u>.
>
> 2) When a river <u>floods</u> onto the flood plain, the water <u>slows down</u> and <u>deposits</u> the <u>eroded material</u> that it's <u>transporting</u>. This <u>builds up</u> the flood plain (makes it <u>higher</u>).
>
> 3) <u>Meanders migrate</u> (move) <u>across</u> the flood plain, making it <u>wider</u>.
>
> 4) The <u>deposition</u> that happens on the <u>slip-off slopes</u> of meanders also <u>builds up</u> the flood plain.

Flood plain

Flood plains and leeves are found in the lower course of a river.

2) <u>Levees</u>

> These are <u>natural embankments</u> (raised bits) along the <u>edges</u> of a <u>river channel</u>. During a flood, <u>eroded material</u> is <u>deposited</u> over the whole flood plain. The <u>heaviest material</u> is <u>deposited closest</u> to the river channel, because it gets <u>dropped first</u> when the river <u>slows down</u>. <u>Over time</u>, the <u>deposited material builds up</u>, creating <u>levees</u> along the edges of the channel, e.g. along the Yellow River in China.

Channel edges | Heavy material deposited during flood | Levees created after repeated flooding

Some river landforms are beautiful — others are gorge-ous...

Step over the <u>hard rock</u> and <u>plunge</u> into the <u>pool</u> — that's how I remember how <u>waterfalls</u> are formed. Geography examiners <u>love river landforms</u> (they're a bit weird like that) so make sure you learn <u>how they form</u>.

Rivers on Maps

You can know all the facts about <u>rivers</u>, but if you don't know what their <u>features</u> look like on <u>maps</u> then some of the exam questions will be a wee bit tricky. Here's something I prepared earlier...

Contour Lines Tell you the Direction a River Flows

<u>Contour lines</u> are the <u>orange lines</u> drawn all over maps. They tell you about the <u>height</u> of the land (in metres) by the numbers marked on them, and the <u>steepness</u> of the land by how <u>close together</u> they are (the <u>closer</u> they are, the <u>steeper</u> the slope).

It sounds obvious, but rivers <u>can't</u> flow uphill. Unless gravity's gone screwy, a river flows <u>from higher</u> contour lines <u>to lower</u> ones. Have a look at this map of Cawfell Beck:

Take a peek at pages 144-145 for more on reading maps.

① The <u>height values</u> get <u>smaller</u> towards the <u>west</u> (left), so west is <u>downhill</u>.

② Cawfell Beck is flowing from <u>east</u> to <u>west</u> (right to left).

③ A <u>V-shape</u> is formed where the contour lines <u>cross</u> the river. The V-shape is <u>pointing uphill</u> to where the river came from.

Maps contain Evidence for River Courses and Landforms

Exam questions might ask you to look at a <u>map</u> and give the <u>evidence</u> for a <u>river course</u> or <u>landform</u>. Learn this stuff and those questions will be a breeze:

Evidence for a river's upper course

The nearby land is <u>high</u> (712 m).

The river <u>crosses lots</u> of <u>contour lines</u> in a <u>short distance</u>, which means it's <u>steep</u>.

The river's <u>narrow</u> (a <u>thin</u> blue line).

The <u>contour lines</u> are very <u>close together</u> and the valley floor is narrow. This means the river is in a <u>steep-sided V-shaped</u> valley.

Evidence for a waterfall

<u>Waterfalls</u> are marked on maps, but the <u>symbol for a cliff</u> (black, blocky lines) and the <u>close contour lines</u> are evidence for a waterfall.

Evidence for a river's lower course

The nearby land is <u>low</u> (less than 20 m).

The river only <u>crosses one contour line</u> so it's <u>very gently sloping</u>.

Another piece of evidence would be the river <u>joining</u> a <u>sea</u> or <u>lake</u>.

The river's <u>wide</u> (a <u>thick</u> blue line).

The river meanders across a large flat area (<u>no contours</u>), which is the <u>flood plain</u>.

The river has <u>large meanders</u>.

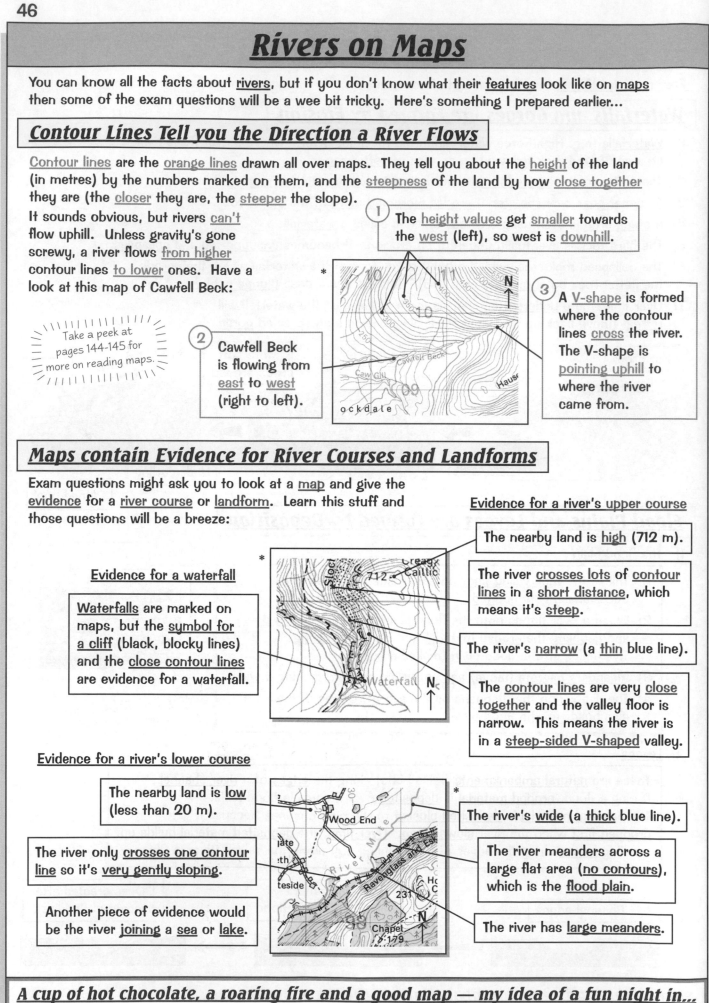

A cup of hot chocolate, a roaring fire and a good map — my idea of a fun night in...

<u>Map</u> questions can be a goldmine of easy marks — all you have to do is <u>say what you see</u>. You just need to understand what the maps are <u>showing</u>, so read this page like there's no tomorrow, then see if you can remember it all.

Unit 1B — Water on the Land

River Discharge

We've not really talked much about the actual <u>water</u> in a river. Well, all that's about to change — hooray.

River Discharge is the Volume of Water Flowing in a River

River discharge is simply the <u>volume of water</u> that flows in a river <u>per second</u>. It's measured in <u>cumecs</u> — cubic metres per second (m³/s). <u>Hydrographs</u> show how the discharge at a <u>certain point</u> in a river <u>changes</u> over time. <u>Storm hydrographs</u> show the changes in river discharge around the time of a <u>storm</u>. Here's an example of a storm hydrograph:

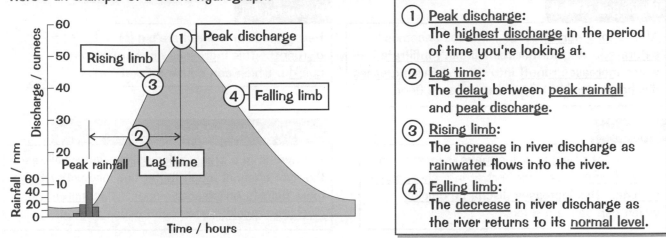

① <u>Peak discharge</u>:
The <u>highest discharge</u> in the period of time you're looking at.

② <u>Lag time</u>:
The <u>delay</u> between <u>peak rainfall</u> and <u>peak discharge</u>.

③ <u>Rising limb</u>:
The <u>increase</u> in river discharge as <u>rainwater</u> flows into the river.

④ <u>Falling limb</u>:
The <u>decrease</u> in river discharge as the river returns to its <u>normal level</u>.

Lag time happens because most rainwater <u>doesn't land directly</u> in the river channel — there's a <u>delay</u> as rainwater <u>gets to the channel</u>. It gets there by <u>flowing quickly overland</u> (called <u>surface runoff</u>, or just <u>runoff</u>), or by <u>soaking into the ground</u> (called <u>infiltration</u>) and flowing <u>slowly underground</u>.

River Discharge is Affected by Different Factors

The <u>more water</u> that flows as <u>runoff</u>, the <u>shorter</u> the <u>lag time</u> will be. This means <u>discharge</u> will increase because <u>more water</u> gets to the channel in a <u>shorter space of time</u>.

Here are a few examples of how different <u>factors</u> can affect discharge:

AMOUNT AND TYPE OF RAINFALL

<u>Lots of rain</u> and <u>short, heavy periods</u> of rainfall means there's <u>more runoff</u>. Lag time is <u>decreased</u>, so discharge <u>increases</u>.

TEMPERATURE

<u>Hot</u>, <u>dry</u> conditions and <u>cold</u>, <u>freezing</u> conditions both result in <u>hard ground</u> — this <u>increases runoff</u>. Lag time is <u>decreased</u>, so discharge <u>increases</u>.

PREVIOUS WEATHER CONDITIONS

After <u>lots of rain</u>, soil can become <u>saturated</u> (it can't absorb any more water). <u>More rainwater</u> won't be able to <u>infiltrate</u> into the soil so <u>runoff will increase</u>. Lag time is <u>decreased</u>, so discharge <u>increases</u>.

ROCK TYPE

Water <u>infiltrates</u> through <u>pore spaces</u> in <u>permeable rock</u> and <u>flows along cracks</u> in <u>pervious rocks</u> — this means there <u>isn't much runoff</u>. Lag time is <u>increased</u>, so discharge <u>decreases</u>.
Water <u>can't</u> infiltrate into <u>impermeable</u> rock — this means there's <u>a lot</u> of <u>runoff</u>. Lag time is <u>decreased</u>, so discharge <u>increases</u>.

LAND USE

<u>Urban areas</u> have <u>drainage systems</u> and they're covered with <u>impermeable materials</u> like concrete — these <u>increase runoff</u>. Lag time is <u>decreased</u>, so discharge <u>increases</u>.

RELIEF (how the height of the land changes)

<u>Lots of runoff</u> occurs on <u>steep slopes</u>. Lag time is <u>decreased</u>, so discharge <u>increases</u>.

Revision lag time — the time between starting and getting bored...

You need to know how <u>different factors</u> can affect the <u>discharge</u> of a river and <u>hydrographs</u> are a helpful way of understanding what's going on. They sound like something from the future, but the excitement ends there I'm afraid.

Flooding

Flooding happens when the <u>level</u> of a river gets <u>so high</u> that it <u>spills over</u> its <u>banks</u>. Sometimes a flood <u>happens without warning</u> — these floods are called <u>flash floods</u>.

Rivers Flood due to Physical Factors

The <u>river level increases</u> when the <u>discharge increases</u> because a high discharge means there's <u>more water in the channel</u>. This means the factors that <u>increase discharge</u> can also <u>cause flooding</u>:

Prolonged rainfall

After a <u>long period</u> of rain, the soil becomes <u>saturated</u>. Any further rainfall <u>can't infiltrate</u>, which <u>increases runoff</u> into rivers. This <u>increases discharge quickly</u>, which can cause a flood.

Heavy rainfall

Heavy rainfall means there's <u>a lot of runoff</u>. This <u>increases discharge quickly</u>, which can cause a flood.

Snowmelt

When a lot of <u>snow</u> or <u>ice melts</u> it means that a <u>lot of water</u> goes into a river in a <u>short space of time</u>. This <u>increases discharge quickly</u>, which can cause a flood.

Relief (how the height of the land changes)

If a river is in a <u>steep-sided valley</u>, water will reach the river channel <u>much faster</u> because water <u>flows more quickly</u> on <u>steeper slopes</u>. This <u>increases discharge quickly</u>, which can cause a flood.

Rivers also Flood because of Human Factors

Here are a couple of examples of how <u>human actions</u> can make flooding <u>more frequent</u> and <u>more severe</u>:

Deforestation

Trees <u>intercept</u> rainwater on their leaves, which then <u>evaporates</u>. Trees also <u>take up water</u> from the ground and <u>store it</u>. This means <u>cutting down</u> trees <u>increases</u> the <u>volume</u> of water that <u>reaches</u> the river channel, which <u>increases discharge</u> and makes flooding <u>more likely</u>.

Building construction

Buildings are often made from <u>impermeable materials</u>, e.g. concrete, and they're surrounded by <u>roads</u> made from <u>tarmac</u> (also impermeable). Impermeable surfaces <u>increase runoff</u> and <u>drains</u> quickly take runoff to rivers. This <u>increases discharge quickly</u>, which can cause a flood.

River Flooding in the UK Appears to be Happening More Often

Some rivers in the UK have been flooding <u>more frequently</u> over the <u>last 20 years</u>. For example, the <u>River Ouse</u> in Yorkshire reached a high water level <u>29 times</u> between 1966 and 1986. But between 1987 and 2007 it reached the same level <u>80 times</u>.

The table shows the <u>locations</u> and <u>dates</u> of some of the big <u>floods</u> that have happened in the UK <u>since 1988</u>.

Year	Rivers	Places affected
1988	Kenwyn	Cornwall
1990	Severn	Gloucestershire
1994	Lavant, Clyde	West Sussex, Glasgow
1998	Severn, Trent, Wye	The Midlands, Mid and South Wales
2000	Ouse, Alyn	Yorkshire, North Wales
2004	Valency	Cornwall
2005	Eden	Cumbria and North Yorkshire
2007	Many	Many parts of the UK
2008	Severn	South Midlands

Lots of water causes flooding — this mainly applies to rivers and toddlers...

If you're having a bath and you leave the taps on the water will eventually <u>go over the sides</u>. It's the same with rivers — they've got a <u>limit</u> to the <u>volume</u> of water they can hold. Please don't experiment by flooding the bathroom though.

Flooding — Case Studies

Everyone's favourite part of geography — the inevitable <u>case studies</u>. Put your learning hats on...

Rich and Poor parts of the World are Affected Differently by Flooding

The <u>effects</u> of floods and the <u>responses</u> to them are <u>different</u> in different parts of the world. A lot depends on how <u>wealthy</u> the part of the world is. Learn the following case studies — you might have to <u>compare</u> two floods like these in your exam:

Flood in a <u>rich</u> part of the world:

<u>Place</u>: Carlisle, England
<u>Date</u>: 8th January, 2005
<u>River</u>: Eden

Flood in a <u>poor</u> part of the world:

<u>Place</u>: South Asia (Bangladesh and India)
<u>Date</u>: July and August, 2007
<u>Rivers</u>: Brahmaputra and Ganges

	Carlisle, England	South Asia (Bangladesh and India)
Causes	• <u>Heavy rainfall</u> — <u>200 mm</u> of rain fell in <u>36 hours</u>. The continuous rainfall <u>saturated</u> the soil, <u>increasing runoff</u> into the River Eden. • Carlisle is a <u>large urban area</u> — <u>impermeable materials</u> like concrete <u>increased runoff</u>. • This caused the <u>discharge</u> of the River Eden to reach <u>1520 cumecs</u> (its <u>average discharge</u> is <u>52 cumecs</u>).	• <u>Heavy rainfall</u> — in one region, <u>900 mm</u> of rain fell in July. The continuous rainfall <u>saturated</u> the soil, <u>increasing runoff</u> into rivers. • <u>Melting snow</u> from glaciers in the <u>Himalayan mountains</u> <u>increased the discharge</u> of the Brahmaputra river. • The <u>peak discharge</u> of <u>both rivers</u> happened <u>at the same time</u>, which <u>increased discharge</u> downstream.
Primary effects	• <u>3 deaths</u>. • Around <u>3000</u> people were made <u>homeless</u>. • <u>4 schools</u> were severely flooded. • <u>350 businesses</u> were shut down. • <u>70 000 addresses</u> lost power. • Some <u>roads</u> and <u>bridges</u> were <u>damaged</u>. • Rivers were <u>polluted</u> with <u>rubbish and sewage</u>.	• Over <u>2000 deaths</u>. • Around <u>25 million</u> people were made <u>homeless</u>. • <u>44 schools</u> were totally <u>destroyed</u>. • Many <u>factories closed</u> and lots of <u>livestock</u> were <u>killed</u>. • <u>112 000 houses</u> were destroyed in India. • <u>10 000 km</u> of <u>roads</u> were destroyed. • Rivers were <u>polluted</u> with <u>rubbish and sewage</u>.
Secondary effects	• Children <u>lost</u> out on <u>education</u> — one school was <u>closed for months</u>. • <u>Stress-related illnesses</u> increased after the floods. • Around <u>3000 jobs</u> were <u>at risk</u> in businesses affected by floods.	• Children <u>lost</u> out on <u>education</u> — around <u>4000 schools</u> were <u>affected</u> by the floods. • Around <u>100 000 people</u> caught <u>water-borne diseases</u> like dysentery and diarrhoea. • Flooded fields <u>reduced</u> basmati rice yields — <u>prices rose 10%</u>. • Many <u>farmers</u> and <u>factory workers</u> became <u>unemployed</u>.
Immediate response	• People were <u>evacuated</u> from areas that flooded. • <u>Reception centres</u> were opened around Carlisle to provide <u>food</u> and <u>drinks</u> for evacuees. • <u>Temporary accommodation</u> was set up for the people made homeless.	• Many people <u>didn't evacuate</u> from areas that flooded, and <u>blocked</u> transport links <u>slowed down</u> any <u>evacuations</u> that were attempted. • <u>Other governments</u> and <u>international charities</u> distributed <u>food</u>, <u>water</u> and <u>medical aid</u>. <u>Technical equipment</u> like <u>rescue boats</u> were also sent to help people who were <u>stranded</u>.
Long-term response	• <u>Community groups</u> were set up to provide <u>emotional support</u> and to give <u>practical help</u> to <u>people</u> who were <u>affected</u> by the floods. • A <u>flood defence scheme</u> has been set up to <u>improve flood defences</u>, e.g. build up <u>banks</u> on the River Eden to <u>prevent flooding</u>.	• <u>International charities</u> have funded the <u>rebuilding of homes</u> and the <u>agriculture</u> and <u>fishing industries</u>. • Some homes have been <u>rebuilt on stilts</u>, so they're <u>less likely</u> to be <u>damaged</u> by future floods.

Flood your mind — with knowledge...

Well, it's pretty clear that floods have <u>different impacts</u> and the <u>responses</u> to them are <u>different</u> when you compare a <u>rich</u> and a <u>poor</u> part of the world. You need to know a couple of <u>examples</u>, so it might as well be these two.

Hard vs Soft Engineering

Floods can be <u>devastating</u>, but there are a number of different <u>strategies</u> to stop them or lessen the blow.

Engineering can Reduce the Risk of Flooding or its Effects

There are <u>two</u> types of strategy to <u>deal with flooding</u>:

<u>Hard engineering</u> — <u>man-made structures</u> built to <u>control the flow</u> of rivers and <u>reduce flooding</u>.

<u>Soft engineering</u> — schemes set up using <u>knowledge</u> of a <u>river</u> and its <u>processes</u> to <u>reduce the effects of flooding</u>.

There's <u>debate</u> about <u>which strategies are best</u>, so you'll need to know the <u>benefits</u> and <u>costs</u> of a few of them. Here's a lovely big table for you to enjoy:

	Method	What it is	Benefits	Disadvantages
HARD ENGINEERING	Dams and reservoirs	<u>Dams</u> (huge walls) are built <u>across</u> the rivers, usually in the <u>upper course</u>. A <u>reservoir</u> (artificial lake) is formed <u>behind</u> the dam.	Reservoirs <u>store water</u>, especially during periods of prolonged or heavy rain, which <u>reduces</u> the <u>risk of flooding</u>. The water in the reservoir is used as <u>drinking water</u> and can be used to <u>generate hydroelectric power</u> (HEP).	Dams are <u>very expensive</u> to build. Creating a reservoir can <u>flood existing settlements</u>. Eroded material is <u>deposited</u> in the <u>reservoir</u> and <u>not</u> along the river's <u>natural course</u> so <u>farmland</u> downstream can be <u>less fertile</u>.
	Channel straightening	The river's <u>course</u> is <u>straightened</u> — <u>meanders</u> are <u>cut out</u> by building <u>artificial straight channels</u>.	Water moves out of the area <u>more quickly</u> because it doesn't travel as far — <u>reducing</u> the <u>risk</u> of flooding.	<u>Flooding</u> may happen <u>downstream</u> of the straightened channel instead, as flood water is <u>carried there faster</u>. There's <u>more erosion downstream</u> because the water's <u>flowing faster</u>.
SOFT ENGINEERING	Flood warnings	The <u>Environment Agency</u> warns people about possible flooding through <u>TV</u>, <u>radio</u>, <u>newspapers</u> and <u>the internet</u>.	The <u>impact</u> of flooding is <u>reduced</u> — warnings give people time to <u>move possessions upstairs</u>, put <u>sandbags</u> in position and to <u>evacuate</u>.	Warnings <u>don't stop</u> a <u>flood</u> from happening. Living in a place that gets <u>lots of warnings</u> could make it <u>difficult</u> to get <u>insurance</u>. People may <u>not</u> hear or have <u>access</u> to warnings.
	Preparation	Buildings are <u>modified</u> to <u>reduce</u> the amount of <u>damage</u> a flood could cause. People make <u>plans</u> for what to do in a flood — they keep <u>important documents</u> and items like <u>torches</u> and <u>blankets</u> in a <u>handy place</u>.	The <u>impact</u> of flooding is <u>reduced</u> — buildings are <u>less damaged</u> and people <u>know what to do</u> when a flood happens. People are also <u>less likely to worry</u> about the threat of floods if they're prepared.	Preparation <u>doesn't guarantee safety</u> from a flood. It could give people a <u>false sense of security</u>. It's <u>expensive</u> to modify homes and businesses.
	Flood plain zoning	Restrictions <u>prevent building</u> on parts of a flood plain that are <u>likely to be affected</u> by a flood.	The <u>risk of flooding</u> is <u>reduced</u> — <u>impermeable surfaces aren't created</u>, e.g. buildings and roads. The <u>impact</u> of flooding is <u>reduced</u> — there aren't any houses or roads to be damaged.	The <u>expansion</u> of an <u>urban area</u> is <u>limited</u> if there aren't any other suitable building sites. It's no help in areas that have <u>already been built on</u>.
	'Do nothing'	<u>No money</u> is spent on <u>new</u> engineering methods or <u>maintaining</u> existing ones. Flooding is a <u>natural process</u> and people should <u>accept the risks</u> of living in an area that's <u>likely to flood</u>.	The river <u>floods</u>, eroded material is <u>deposited</u> on the flood plain, making <u>farmland more fertile</u>.	The <u>risk</u> of flooding and the <u>impacts</u> of flooding <u>aren't reduced</u>. A flood will probably cause <u>a lot of damage</u>.

No, river straightening isn't done with a gigantic pair of hot ceramic plates...

I wasn't lying — the table was both lovely and big. It looks like a lot of stuff to learn, but it's not difficult at all. Make sure you have at least a couple of <u>benefits</u> and <u>disadvantages</u> for <u>each method</u> stashed away in your brain.

Managing the UK's Water

You think you can't live without your MP3 player, but believe me, it's <u>water</u> you <u>can't live without</u>.

The Demand for Water is Different Across the UK

In the UK, the places with a <u>good supply</u> of water <u>aren't the same</u> as the places with the <u>highest demand</u>:

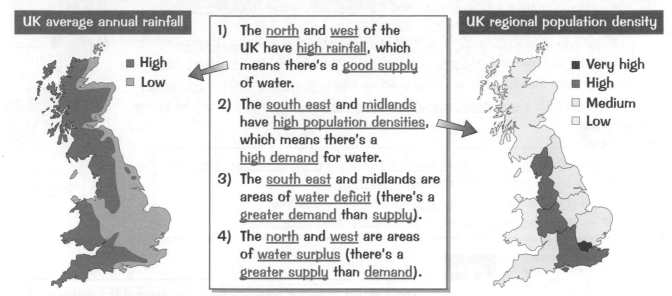

UK average annual rainfall
- ■ High
- □ Low

1) The <u>north</u> and <u>west</u> of the UK have <u>high rainfall</u>, which means there's a <u>good supply</u> of water.

2) The <u>south east</u> and <u>midlands</u> have <u>high population densities</u>, which means there's a <u>high demand</u> for water.

3) The <u>south east</u> and midlands are areas of <u>water deficit</u> (there's a <u>greater demand</u> than <u>supply</u>).

4) The <u>north</u> and <u>west</u> are areas of <u>water surplus</u> (there's a <u>greater supply</u> than <u>demand</u>).

UK regional population density
- ■ Very high
- ■ High
- □ Medium
- □ Low

The <u>demand</u> for water in the UK is <u>increasing</u>:

1) Over the <u>past 25 years</u>, the amount of water <u>used</u> by people in the UK has <u>gone up</u> by about <u>50%</u>.

2) The <u>UK population</u> is predicted to <u>increase</u> by around <u>10 million</u> people over the next <u>20 years</u>.

The UK needs to Manage its Supply of Water...

1) One way to deal with the <u>supply and demand problem</u> is to <u>transfer water</u> from areas of <u>surplus</u> to areas of <u>deficit</u>. For example, <u>Birmingham</u> (an area of <u>deficit</u>) is supplied with water from the <u>middle of Wales</u> (an area of <u>surplus</u>).

2) Water transfer can cause a variety of <u>issues</u>:

- The <u>dams</u> and <u>aqueducts</u> (bridges used to <u>transport water</u>) that are needed are <u>expensive</u>.
- It could <u>affect the wildlife</u> that lives in the rivers, e.g. <u>fish migration</u> patterns could be disrupted by dam building.
- There might be <u>political issues</u>, e.g. people <u>may not want</u> their <u>water given to another country</u>.

3) Another way to <u>increase water supplies</u> in deficit areas is to build <u>more reservoirs</u> to <u>store more water</u>. However, building a reservoir can involve <u>flooding settlements</u> and <u>relocating people</u>.

4) <u>Fixing leaky pipes</u> would mean <u>less water</u> is <u>lost</u> during transfer. For example, <u>millions of litres</u> of water are <u>lost everyday</u> through leaky pipes around <u>London</u> — fixing leaky pipes would save some of this.

...and Reduce its Demands for Water

1) People can <u>reduce</u> the amount of water that they <u>use</u> at home, e.g. by taking showers instead of baths, running washing machines only when they're full and by using hosepipes less.

2) <u>Water companies</u> want people to have <u>water meters</u> installed — meters are used to <u>charge people</u> for the <u>exact volume</u> of water that they use. People with water meters are more likely to be <u>careful</u> with the <u>amount</u> of water they use — they're <u>paying for every drop</u>.

I'd like a glass of water please — hang on, I'll just nip up to Scotland...

After reading this topic you probably won't be leaving the tap on while you brush your teeth again. The UK isn't a desert by any means, but you need to know about the <u>increasing demand for water</u> and how <u>supplies</u> can be <u>managed</u>.

UK Reservoir — Case Study

Joy of joys, another <u>case study</u>. It's about a <u>reservoir</u> in the UK that <u>supplies water</u> to a lot of people...

Rutland Water is a Reservoir in the East Midlands

1) The dam was built and Rutland Water was created during the <u>1970s</u>.

2) The reservoir covers a <u>12 km²</u> area and it's <u>filled</u> with water from <u>two rivers</u> — the <u>River Welland</u> and the <u>River Nene</u>.

3) Rutland Water was designed to <u>supply</u> the <u>East Midlands</u> with <u>more water</u> — enough to cope with <u>rapid population growth</u> in places like <u>Peterborough</u>.

4) Areas around the reservoir are also used as a <u>nature reserve</u> and for <u>recreation</u>.

Here are some of the <u>economic</u>, <u>social</u> and <u>environmental</u> impacts of Rutland Water:

Rutland Water

Peterborough

Economic

- The reservoir <u>boosts</u> the <u>local economy</u> — it's a <u>popular tourist attraction</u> because of the <u>wildlife</u> and <u>recreation facilities</u>.

- Around <u>6 km²</u> of <u>land</u> was <u>flooded</u> to create the reservoir. This included <u>farmland</u>, so some <u>farmers lost their livelihoods</u>.

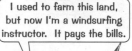
I used to farm this land, but now I'm a windsurfing instructor. It pays the bills.

Social

- Lots of <u>recreational activities</u> take place <u>on</u> and <u>around</u> the reservoir, e.g. sailing, windsurfing, birdwatching and cycling.

- Many <u>jobs</u> have been <u>created</u> to <u>build</u> and <u>maintain</u> the <u>reservoir</u>, and to <u>run</u> the <u>nature reserve</u> and <u>recreational activities</u>.

- Schools use the reservoir for <u>educational visits</u>.

- <u>Two villages</u> were <u>demolished</u> to make way for the reservoir.

Environmental

- Rutland Water is a <u>Site of Special Scientific Interest</u> (<u>SSSI</u>) — an area where wildlife is protected.

- <u>Hundreds</u> of <u>species</u> of <u>birds</u> live around the reservoir and <u>tens of thousands</u> of <u>waterfowl</u> (birds that live on or near water) come to Rutland Water over the <u>winter</u>.

- A <u>variety of habitats</u> are found around the reservoir, e.g. marshes, mudflats and lagoons. This means <u>lots of different organisms</u> live in or around the reservoir.

- <u>Ospreys</u> (fish-eating birds of prey that were extinct in Britain) have been <u>reintroduced</u> to central England by the <u>Rutland Osprey Project</u> at the reservoir.

- A <u>large area</u> of land was <u>flooded</u> to <u>create</u> the reservoir, which <u>destroyed some habitats</u>.

Rutland Water has to be Managed Sustainably

The <u>supply of water</u> from the reservoir has to be <u>sustainable</u>. This means that people should be able to get all the water they need <u>today</u>, <u>without stopping</u> people in the <u>future</u> from having <u>enough water</u>.

Basically, people today <u>can't deplete</u> the <u>water supply</u> or <u>damage the environment</u> too much, or the supply won't be the same in the <u>future</u>. To use the reservoir in a <u>sustainable way</u>, people can <u>only take out</u> as much water as is <u>replaced</u> by the rivers that supply it. That way, the supply will <u>stay the same</u> for the future.

Don't get stuck in a Rut-land while doing your revision...

Congratulations, another topic under your belt — your prize awaits you on the next page. OK, so it's a <u>revision</u> <u>summary</u>, but the real prize is the satisfaction of knowing everything about <u>rivers</u>. I hope that sounded convincing...

Revision Summary for Water on the Land

Wat-er load of fun that was. Now it's time to see how much information your brain has soaked up. I think you'll be surprised — I reckon about 16 litres of knowledge has been taken in. Have a go at the questions below and then go back over the topic to check your answers. If something's not quite right, pore over the page again. Once you can answer everything correctly you're ready to sail away, sail away, sail away... to the next topic.

1) What does a river's long profile show?
2) Describe the cross profile of a river's lower course.
3) Name the river course where vertical erosion is dominant.
4) What's the difference between abrasion and attrition?
5) Name two processes of transportation.
6) When does deposition occur?
7) Where is the current fastest on a meander?
8) Name the landform created when a meander is cut off by deposition.
9) Where do waterfalls form?
10) How is a gorge formed?
11) What is a flood plain?
12) Describe how levees are formed.
13) What do the contour lines on a map show?
14) Give two pieces of map evidence for a waterfall.
15) Give two pieces of map evidence for a river's lower course.
16) What is river discharge?
17) How does impermeable rock affect river discharge?
18) Describe two physical factors that can cause floods.
19) Describe one human factor that can cause floods.
20) a) Name a flood that happened in a rich part of the world.
    b) Describe the primary effects of the flood.
    c) Describe the long-term responses to the flood.
21) Give an example of a flood in a poor part of the world.
22) Define hard engineering.
23) Define soft engineering.
24) Describe how channel straightening reduces the risk of a flood.
25) Describe the disadvantages of flood warnings.
26) Which areas of the UK have a water deficit?
27) Give one potential problem of water transfer.
28) a) Name a reservoir in the UK.
    b) Describe two social impacts that the reservoir has had.
    c) Describe two environmental impacts that the reservoir has had.
29) What is meant by a sustainable water supply?
30) How can water be taken from a reservoir in a sustainable way?

Ice Levels Over Time

It's time to put your hat, scarf and jumper on, you've arrived at the topic that's all about <u>ice</u>...

The Earth has Glacial Periods and Interglacial Periods

1) The Earth goes through <u>cold periods</u> which last for <u>millions of years</u> called <u>ice ages</u>. During ice ages, <u>large masses of ice</u> cover parts of the Earth's surface.

2) The last <u>ice age</u> was the <u>Pleistocene</u> that <u>began</u> around <u>2.6 million years ago</u>.

3) During ice ages there are <u>cooler periods</u> called <u>glacial periods</u> when the ice <u>advances</u> to cover <u>more</u> of the Earth's surface. Each one lasts for about <u>100 000 years</u>.

4) <u>In between</u> the glacial periods are <u>warmer periods</u> called <u>interglacial periods</u> when the ice <u>retreats</u> to cover <u>less</u> of the Earth's surface. Each one lasts around <u>10 000 years</u>.

5) The <u>last glacial period</u> began around <u>100 000 years ago</u> and ended around <u>10 000 years ago</u>.

Ice Covered Much More of the Earth's Surface 20 000 Years Ago

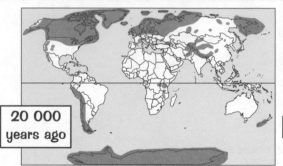

20 000 years ago

Today

■ = Ice

1) Since the beginning of the <u>Pleistocene</u> there have been <u>permanent ice sheets</u> on <u>Greenland</u> and <u>Antarctica</u>. Ice has also covered <u>other parts of the world</u> during the <u>colder glacial periods</u>.

2) Ice covered a lot more of the land around <u>20 000 years ago</u> (during the <u>last glacial period</u>) — over <u>30%</u> of the Earth's land surface was <u>covered by ice</u>, including <u>nearly all</u> of <u>the UK</u>.

3) We're <u>currently</u> in an <u>interglacial period</u> that <u>began</u> around <u>10 000 years ago</u>. Today about <u>10%</u> of the Earth's land surface is <u>covered by ice</u> — the <u>only ice sheets</u> are the ones on <u>Greenland</u> and <u>Antarctica</u>.

> Ice sheets are huge masses of ice that cover whole continents. Glaciers are masses of ice which fill valleys and hollows.

Evidence of Changing Temperature Comes From Three Main Sources

CHEMICAL EVIDENCE

The <u>chemical composition</u> of <u>ice</u> and <u>marine sediments</u> change as temperature changes, so they can be used to work out how <u>global temperature</u> has <u>changed</u> in the past. Ice and sediments build up over thousands of years so <u>samples</u> taken at <u>different depths</u> show the temperature over <u>thousands of years</u>. The records show a <u>pattern</u> of <u>increasing</u> and <u>decreasing temperature</u>, which caused the ice to <u>advance</u> and <u>retreat</u>.

GEOLOGICAL EVIDENCE

Some <u>landforms</u> we can see <u>today</u> were created by glaciers in the <u>past</u> (see p. 57). This shows that some <u>areas</u> that <u>aren't covered in ice</u> today were <u>covered in the past</u>, which means temperatures were <u>lower</u>.

FOSSIL EVIDENCE

The <u>remains</u> of some organisms are <u>preserved</u> when they die, creating <u>fossils</u>. Fossils show the <u>distribution</u> of plants and animals that are <u>adapted</u> to <u>warm</u> or <u>cold</u> <u>climates</u> at different times in the past. From this we can tell which <u>areas</u> were <u>warmer</u> or <u>colder</u> in the past.

Pleistocene — comes in many colours and gets stuck in the carpet...

Don't take your warm clothes off yet — there's lots more <u>ice</u> to come. Make sure you know the difference between an <u>ice age</u> and a <u>glacial period</u>, when the last glacial period was, and how far the ice sheets <u>spread</u> in that period.

Glacial Budget

Glaciers are masses of ice that fill valleys and hollows. They move downhill under the force of gravity. They also have a budget — £10 a week on ice cream, £5 on glacier mints and £2 on freezer pops.

A Glacier has a Zone of Accumulation and a Zone of Ablation

1) Accumulation is the input of snow and ice into the glacier.

2) Ablation is the output of water from a glacier as the ice melts.

INPUTS

3) You get more accumulation than ablation in the upper part of a glacier — so it's called the zone of accumulation.

4) You get more ablation than accumulation in the lower part of a glacier — so it's called the zone of ablation.

OUTPUTS

It can look like glaciers stay in the same place, but remember — they're melting at the bottom end, growing at the top end and always moving like a conveyer belt.

The Difference Between Accumulation and Ablation is the Glacial Budget

1) The glacial budget is the difference between total accumulation and total ablation for one year.

2) The amount of ice in a glacier, and whether it's advancing or retreating, depends on the glacial budget:
- A positive glacial budget is when accumulation (input) exceeds ablation (output). The glacier gets larger and the snout (the bottom end of the glacier) advances down the valley.
- A negative glacial budget is when ablation (output) exceeds accumulation (input) The glacier gets smaller and the snout retreats up the valley.
- If there's the same amount of accumulation and ablation over a year, the glacier stays the same size and the position of the snout doesn't change.

The Glacial Budget Changes in the Short-Term and the Long-Term

Temperature changes throughout each year and over the years. Both these things affect the glacial budget:

1) Glaciers advance and retreat seasonally:
- In the summer there's more ablation than accumulation because more ice melts when it's warm. This means there's a negative glacial budget so glaciers retreat.
- In the winter there's more accumulation than ablation because there's more snowfall and less melting. This means there's a positive glacial budget, so glaciers advance.

2) Since 1950 most glaciers have had a negative glacial budget, so they've been retreating. This is because the earth's temperature has been increasing (global warming).

Glacial budget for a glacier in the northern hemisphere

Winter — positive glacial budget | Summer — negative glacial budget

Volume of water

Ablation | Accumulation

October | January | April | July | October

A glacial budget — it's all about having a n-ice time on the cheap...

There are a few technical terms to learn here — just make sure you know how accumulation and ablation are linked to the glacial budget and you'll be fine in the exam. It'll help if you can scribble simple versions of the diagrams too.

Glacier — Case Study

If you fancy a sight-seeing trip to a glacier you'd better book it quick. <u>Most glaciers</u> are <u>getting smaller</u>, some by up to a few metres each year — and some are in serious danger of <u>disappearing completely</u>.

The Rhône Glacier is Retreating

1) The <u>Rhône Glacier</u> is in the <u>Swiss Alps</u>.

2) It's currently about <u>7.8 km long</u>.

3) Like <u>most</u> of the <u>glaciers</u> in the world, it's been <u>retreating</u> since the <u>19th century</u>.

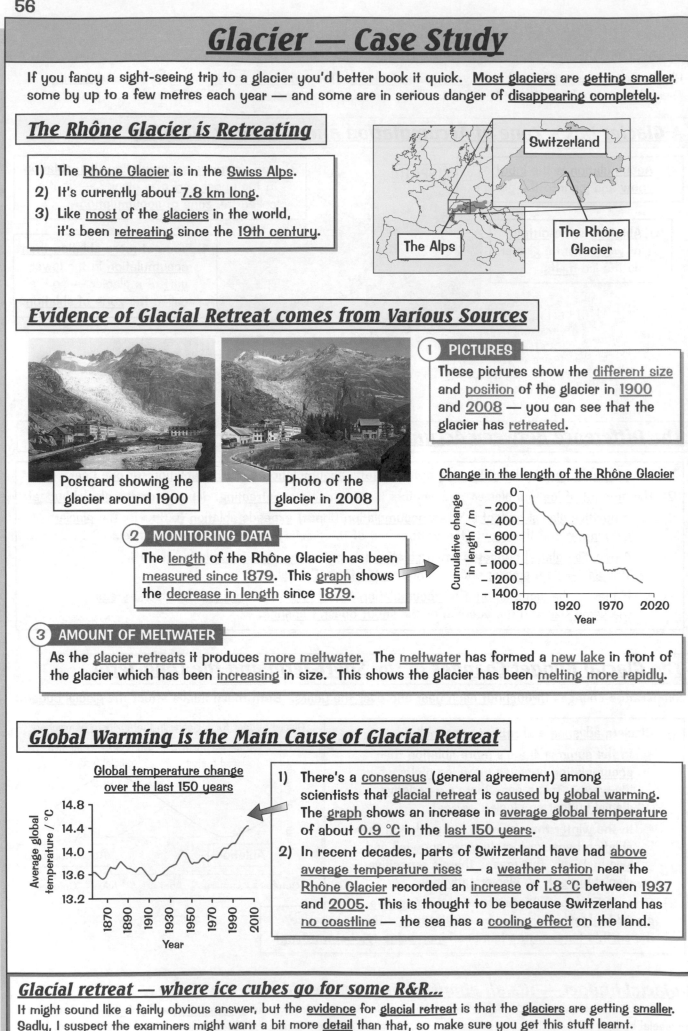

Switzerland

The Alps

The Rhône Glacier

Evidence of Glacial Retreat comes from Various Sources

Postcard showing the glacier around 1900

Photo of the glacier in 2008

1 PICTURES

These pictures show the <u>different size</u> and <u>position</u> of the glacier in <u>1900</u> and <u>2008</u> — you can see that the glacier has <u>retreated</u>.

2 MONITORING DATA

The <u>length</u> of the Rhône Glacier has been <u>measured since 1879</u>. This <u>graph</u> shows the <u>decrease in length</u> since <u>1879</u>.

Change in the length of the Rhône Glacier

Cumulative change in length / m

3 AMOUNT OF MELTWATER

As the <u>glacier retreats</u> it produces <u>more meltwater</u>. The <u>meltwater</u> has formed a <u>new lake</u> in front of the glacier which has been <u>increasing</u> in size. This shows the glacier has been <u>melting more rapidly</u>.

Global Warming is the Main Cause of Glacial Retreat

Global temperature change over the last 150 years

Average global temperature / °C

1) There's a <u>consensus</u> (general agreement) among scientists that <u>glacial retreat</u> is <u>caused</u> by <u>global warming</u>. The <u>graph</u> shows an increase in <u>average global temperature</u> of about <u>0.9 °C</u> in the <u>last 150 years</u>.

2) In recent decades, parts of Switzerland have had <u>above average temperature rises</u> — a <u>weather station</u> near the Rhône Glacier recorded an <u>increase</u> of <u>1.8 °C</u> between <u>1937</u> and <u>2005</u>. This is thought to be because Switzerland has <u>no coastline</u> — the sea has a <u>cooling effect</u> on the land.

Glacial retreat — where ice cubes go for some R&R...

It might sound like a fairly obvious answer, but the <u>evidence</u> for <u>glacial retreat</u> is that the <u>glaciers</u> are getting <u>smaller</u>. Sadly, I suspect the examiners might want a bit more <u>detail</u> than that, so make sure you get this stuff learnt.

Glacial Erosion

It might not seem like <u>glaciers</u> do much — after all they're just large blocks of <u>ice</u> sitting around — but they actually cause rather a lot of <u>erosion</u> and have a massive effect on the <u>landscape</u> around us.

Glaciers Erode the Landscape as They Move

1) The <u>weight</u> of the <u>ice</u> in a glacier makes it <u>move downhill</u> (advance), <u>eroding</u> the <u>landscape</u> as it goes.

2) The moving ice <u>erodes</u> the landscape in <u>two</u> ways:

- <u>Plucking</u> occurs when <u>meltwater</u> at the <u>base</u>, <u>back</u> or <u>sides</u> of a glacier <u>freezes onto</u> the <u>rock</u>. As the glacier <u>moves forward</u> it <u>pulls pieces of rock out</u>.

- <u>Abrasion</u> is where <u>bits of rock</u> stuck in the ice <u>grind against</u> the rock below the glacier, <u>wearing it away</u> (it's a bit like the glacier's got sandpaper on the bottom of it).

3) At the top end of the glacier the ice <u>doesn't</u> move in a <u>straight line</u> — it moves in a <u>circular motion</u> called <u>rotational slip</u>. This can erode <u>hollows</u> in the landscape and <u>deepen</u> them into <u>bowl shapes</u>.

4) The rock above glaciers is also <u>weathered</u> (<u>broken down</u> where it is) by the <u>conditions around glaciers</u>. <u>Freeze-thaw weathering</u> is where <u>water</u> gets into <u>cracks</u> in rocks. The water <u>freezes</u> and <u>expands</u>, putting <u>pressure</u> on the rock. The ice then <u>thaws</u>, <u>releasing</u> the pressure. If this process is <u>repeated</u> it can make bits of the rock <u>fall off</u>.

Glacial Erosion Produces Seven Different Landforms

An <u>arête</u> is a <u>steep-sided ridge</u> formed when <u>two</u> glaciers flow in <u>parallel valleys</u>. The glaciers erode the <u>sides</u> of the valleys, which <u>sharpens</u> the <u>ridge between them</u>. (E.g. Striding Edge, Lake District)

A <u>pyramidal peak</u> is a <u>pointed</u> mountain peak with at least <u>three sides</u>. It's formed when <u>three or more</u> back-to-back glaciers <u>erode</u> a mountain. (E.g. Snowdon, Wales)

<u>Truncated spurs</u> are cliff-like edges on the valley side formed when <u>ridges</u> of land (spurs) that stick out into the main valley are <u>cut off</u> as the glacier moves past.

<u>Corries</u> begin as hollows containing a small glacier. As the ice moves by <u>rotational slip</u>, it <u>erodes</u> the hollow into a steep-sided, <u>armchair shape</u> with a lip at the bottom end. When the ice melts it can leave a small circular lake called a <u>tarn</u>. (E.g. Red Tarn, Lake District)

<u>Hanging valleys</u> are valleys formed by <u>smaller glaciers</u> (called <u>tributary glaciers</u>) that flow into the <u>main glacier</u>. The glacial trough is eroded much <u>more deeply</u> by the <u>larger glacier</u>, so when the glaciers melt the valleys are left at a <u>higher level</u>.

<u>Ribbon lakes</u> are <u>long, thin lakes</u> that form after a <u>glacier retreats</u>. They form in <u>hollows</u> where <u>softer rock</u> was <u>eroded more</u> than the surrounding hard rock. (E.g. Windermere, Lake District)

<u>Glacial troughs</u> are <u>steep-sided</u> valleys with <u>flat bottoms</u>. They start off as a <u>V-shaped</u> river valley but change to a <u>U-shape</u> as the glacier erodes the sides and bottom, making it <u>deeper</u> and <u>wider</u>. (E.g. Nant Ffrancon, Snowdonia)

Truncated spurs — when Tottenham finish their match early...

Knowing all these <u>erosional landforms</u> will help you sound clever next time you're up a mountain, oh yeah, and it'll help in the exam too. You might be asked to spot them on a <u>map</u> in the exam — see page 59 for help with that.

Unit 1B — Ice on the Land

Glacial Transport and Deposition

Glaciers are a bit like small children — they do an excellent job of damaging their surroundings and then they leave lots of mess behind. Problem is, no-one's going to clean up after them, the mucky pups.

Glaciers Transport and Deposit Material

1) Glaciers can move material (such as rocks and earth) over very large distances — this is called transportation.

2) The material is frozen in the glacier, carried on its surface, or pushed in front of it. It's called bulldozing when the ice pushes loose material in front of it.

3) When the ice carrying the material melts, the material is dropped on the valley floor — this is called deposition. It also occurs when the ice is overloaded with material.

4) The dropped material makes landforms such as moraines and drumlins (see below).

5) Glacial deposits aren't sorted by weight like river deposits — rocks of all shapes and sizes are mixed up together.

Glaciers Deposit Material as Different Types of Moraine

Moraines are landforms made out of material dropped by a glacier as it melts. There are four different types, depending on their position:

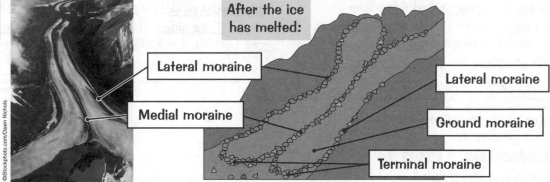

Before the ice melts:

After the ice has melted:

Lateral moraine

Medial moraine

Lateral moraine

Ground moraine

Terminal moraine

©iStockphoto.com/Dawn Nichols

1) Lateral moraine is a long mound of material deposited where the side of the glacier was.

2) Medial moraine is a long mound of material deposited in the centre of a valley where two glaciers met (the two lateral moraines join together).

3) Terminal moraine builds up at the snout of the glacier when it remains stationary. It's deposited as semicircular mounds.

4) Ground moraine is a thin layer of material deposited over a large area as a glacier melts.

Material can also be Deposited as Drumlins

1) Drumlins are elongated hills of glacial deposits — the largest ones can be over 1000 m long, 500 m wide and 50 m high.

2) They're round, blunt and steep at the upstream end, and tapered, pointed and gently sloping at the downstream end.

3) An example of where drumlins can be found is the Ribble Valley, Lancashire.

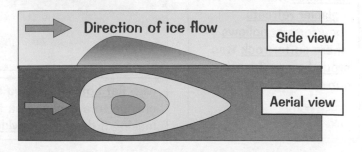

Direction of ice flow

Side view

Aerial view

The glacial weather forecast — cold in most places, with mor-raine on the way...

Check you know the differences between the four different types of moraine — try drawing a simple version of the diagram. If you remember everything on this page you'll be laughing in the exam. Or at least sniggering quietly.

Glacial Landforms on Maps

In the exam you might be asked to spot glacial landforms on an OS® map. It's no problem when you know how, so here are a few tips for you. For a bit more on using maps, turn to pages 144 and 145.

Use Contour Lines to Spot Pyramidal Peaks, Corries and Arêtes on a Map

Contour lines are the orange lines drawn all over maps. They tell you about the height of the land by the numbers marked on them, and the steepness of the land by how close together the lines are (the closer they are, the steeper the slope). Here are a few tips on how to spot pyramidal peaks, arêtes and corries on a map:

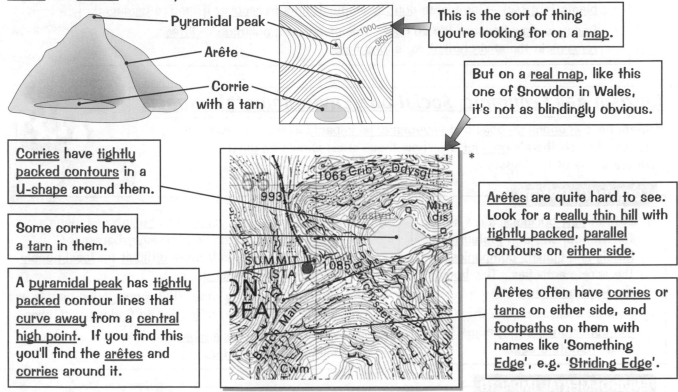

Pyramidal peak

Arête

Corrie
with a tarn

This is the sort of thing you're looking for on a map.

But on a real map, like this one of Snowdon in Wales, it's not as blindingly obvious.

Corries have tightly packed contours in a U-shape around them.

Some corries have a tarn in them.

A pyramidal peak has tightly packed contour lines that curve away from a central high point. If you find this you'll find the arêtes and corries around it.

Arêtes are quite hard to see. Look for a really thin hill with tightly packed, parallel contours on either side.

Arêtes often have corries or tarns on either side, and footpaths on them with names like 'Something Edge', e.g. 'Striding Edge'.

You can also use Maps to Spot Glacial Troughs and Ribbon Lakes

You might be asked to spot a glacial trough or a ribbon lake on a map extract. This map of Nant Ffrancon (a glacial trough) and Llyn Ogwen (a ribbon lake) in Wales shows the classic things to look out for:

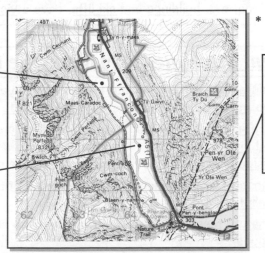

Glacial troughs are flat valleys with very steep sides. There are no contour lines on the bottom of the valley but they're tightly packed on the sides.

Look for a wide, straight valley in a mountainous area with a river that looks too small to have formed the valley.

Many glacial troughs have ribbon lakes in them. Look for a flat valley with steep sides surrounding a long straight lake.

Con-tours — when the Prison Service goes sight-seeing...

There's lots of potential here for easy marks in the exam, just study the map carefully and say what you see. Make sure you refer to the map in your answer though — for help with this, e.g. using grid references, see p. 144.

Impacts and Management of Tourism On Ice

We may think skiing holidays are all about lazy days on the slopes and après-ski indulgence, but they have other impacts apart from shrinking our bank balances, and these impacts need to be managed.

Areas Covered in Snow and Ice are Fragile Environments

Areas that are covered in snow and ice attract lots of tourists for things like winter sports and sightseeing of glaciers. The environments are fragile though — they're easily damaged and difficult to manage:

- There's only a short growing season (when there's enough light and warmth for plants to grow) — so plants don't have much time to recover if they're damaged.
- Decay is slow because it's so cold. This means any pollution or litter remains in the environment for a long time.

There's going to be an impact alright.

Tourism has Economic, Social and Environmental Impacts

Tourists have economic, social and environmental impacts in areas covered in snow and ice, so there's conflict over how these areas should be used. Here are a few of the impacts:

ECONOMIC IMPACTS

1) Lots of new businesses are set up for the tourists, e.g. restaurants, hotels and guiding companies for the sports activities. This boosts the local economy.

2) New businesses means there are job opportunities for people in these remote areas.

SOCIAL IMPACTS

1) Increased numbers of people and businesses mean the infrastructure (roads and railways) becomes congested. This makes it more difficult for local people and tourists to get around.

2) More job opportunities mean that more young people will stay in the area instead of leaving to find work in cities.

3) Tourists can trigger avalanches on ski slopes which can cause injuries and deaths.

ENVIRONMENTAL IMPACTS

1) The fragile glacial environment is damaged by people trampling on the snow and the soil beneath, which causes soil erosion.

2) Glacial landforms like moraines are eroded by people walking on them.

3) There's increased noise, pollution and litter from all the people and traffic in the area.

4) The developments in the area, e.g. buildings and ski lifts, have a visual impact on the environment.

There are Management Strategies to Manage the Different Impacts

There's a need to conserve the fragile environment, but people also have the right to see and experience it. There are different strategies to manage the environment so the impacts of tourism are reduced:

1) Tourists are kept informed of avalanche risks, so they know which areas to avoid. Resorts can build structures to slow and divert the moving snow, plant trees to act as barriers, and set off controlled avalanches to dislodge snow before tourists arrive on the slopes in the morning.

2) Improvements to public transport systems can reduce the amount of traffic and so reduce damage to the environment from pollution.

3) Areas can be set aside as nature reserves. Tourist activity in these areas is limited, so their environmental impact is reduced.

Av-a-lanche break — what Cockney skiers do at midday...

Lots more impacts to get through here. Repeat after me... economic, social and environmental. At least there aren't any political ones for you to remember. Once you know the impacts, check you know how they're managed.

Tourism On Ice — Case Study

So, where to go on holiday... Well, if you're allergic to sea water and know that getting sand in your socks leads to a world of pain, I have a tip for you — rumour has it that Chamonix is nice this time of year.

People go to Chamonix for Winter Sports and Sightseeing

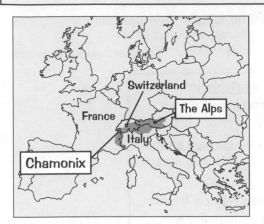

1) The Chamonix Valley is in eastern France at the foot of Mont Blanc (the highest mountain in the Alps). It's close to the border with Italy and Switzerland.

2) It's one of the most popular tourist destinations in the world with around 5 million visitors a year.

3) The region has lots of glaciers, including the Mer de Glace. The Mer de Glace is the longest glacier in France — it's 7 km long and 200 m deep.

4) There are also many other tourist attractions such as 6 ski areas, 350 km of hiking trails, 40 km of mountain bike tracks, an Alpine museum and an exhibition centre.

Tourism has Economic, Social and Environmental Impacts on the Region

ECONOMIC IMPACTS

1) The tourism industry in Chamonix creates a lot of jobs, e.g. 2500 people work as seasonal workers every year.

2) Companies make a lot of money from tourism in Chamonix, e.g. Compagnie du Mont Blanc is a company that runs ski lifts and rail transport — it has a turnover of €50 million.

SOCIAL IMPACTS

1) The types of jobs available in Chamonix have changed from farm labouring to jobs in restaurants and hotels etc.

2) Tourist developments, e.g. ski slopes, have increased the risk of avalanches. This means there are more deaths from avalanches, e.g. in 1999 an avalanche killed 12 people.

ENVIRONMENTAL IMPACTS

1) Large numbers of tourists cause a lot of traffic, which increases pollution. E.g. a study from 2002 to 2004 showed that traffic pollution was worse in the Chamonix region than in the centre of Paris.

2) A huge amount of energy is used to run the facilities for tourists, e.g. the hotels, ski lifts and snow-making machines. This increases CO_2 emissions, which increases global warming.

Tourism in the Resort has to be Carefully Managed

Management of the Chamonix Valley has to balance the need to conserve the environment with the right of people to see and experience it. Here are a few of the strategies used:

1) A system of avalanche barriers is maintained around the resorts, e.g. there's a barrier at Taconnaz. There are also avalanche awareness courses and daily bulletins to keep tourists aware of the risks. This means tourists are less likely to be hurt or killed by an avalanche.

2) The amount of traffic in Chamonix is managed by providing free public transport for tourists. The amount of pollution from public transport is reduced by using low emission buses.

3) Some hotels are reducing their energy use, e.g. by installing solar panels to heat water and systems to automatically turn lights off. This means CO_2 emissions are reduced.

Tourism on Ice — sounds like another piece of TV gold...

Another case study, another day — but at least this one makes you think of holidays. Maybe not the most exciting thing to learn but examiners love to read about 'real world' examples — so the more detail you can shove in the better.

Impacts of Glacial Retreat

The end of this topic is near, but don't slack off yet — there's still one page left. Once it's done though, you'll be an expert on all things glacial. So, wave goodbye with a page on the impacts of glacial retreat.

Glacial Retreat and Unreliable Snowfall Affects Tourism

The economies of many areas that are covered in snow and ice rely on money from tourism (e.g. for winter sports and sightseeing of glaciers). These areas are being affected by glacial retreat (see p. 55) and unreliable snowfall:

1) Glacial retreat means the ice will no longer be available for winter sports, e.g. trekking and ice climbing, or sightseeing of glaciers. This means the area will attract fewer tourists.

2) Unreliable snowfall means that there might not be enough snow for winter sports, e.g. skiing and snowboarding. This also means the area will attract fewer tourists.

3) Fewer tourists will mean that the businesses that rely on tourism, e.g. hotels, restaurants and guiding companies, will make less money and may go out of business.

4) This would lead to increased unemployment in these areas.

Retreat!

Retreat! Retreat!

Glacial Retreat has other Impacts

ECONOMIC IMPACTS

Once a glacier has completely melted, the amount of meltwater decreases. This means industries that rely on the supply of meltwater, e.g. agriculture for irrigation and hydroelectric power (HEP) for electricity production, will make less money and could shut down.

SOCIAL IMPACTS

1) Glacial retreat will mean the water supply to some settlements is reduced (see above).

2) Disruptions to power supplies from HEP could leave some people with an unreliable power supply.

3) If businesses shut down, local people will have to move away to find work. Young people in particular will move away, so older family members might be left behind.

4) If an area's population declines, local services and recreational facilities will also shut down.

5) The ice will no longer be available for recreational use for local people, e.g. for trekking and ice climbing.

ENVIRONMENTAL IMPACTS

1) Glacial retreat is linked to an increase in natural hazards — rapid melting can cause flooding, rockslides and avalanches. These hazards destroy habitats and disrupt food chains.

2) Meltwater from retreating glaciers contributes to rising sea level — water is no longer stored as ice on land and returns to the sea. Rising sea level destroys coastal habitats by causing flooding and erosion.

3) Lots of fish species are adapted to live in the cold meltwater that comes from glaciers. When glaciers have completely melted, there's no cold meltwater so these fish species may die out.

4) Harmful pollutants can be trapped in glacial ice, e.g. the pesticide DDT that was used from the 1940s to 1980s. Rapid melting releases them back into the environment, polluting streams and lakes.

Glacial retreat — don't blame the killer whales, they make the grey seals retreat...

Yep, you've done it, you've seen off the last chilly, ice-related revision page. When you've finished your celebration of choice slide over to the next page and check it's all firmly stuck in your head. Then you can move on to warmer topics.

Revision Summary for Ice on the Land

Time to whip off your hat and scarf and warm up with some revision questions. The bad news is that if you don't know the answers you're going to have to dip back into the icy depths of this topic. Once you're confident of all the answers I recommend that you indulge in some kind of recreational activity. Don't disappear for too long though, there's plenty more revision where this topic came from.

1) What's the name of the last ice age?

2) How long ago did the last glacial period end?

3) How much of the Earth's land surface is currently covered by ice?

4) What three types of evidence are used to identify past temperature changes?

5) Define the term accumulation.

6) What is the zone of ablation?

7) What happens when a glacier has a positive glacial budget?

8) Why does a glacier retreat in summer?

9) a) Give an example of a retreating glacier.

   b) What evidence is there that the glacier you named has retreated since the 19th century?

   c) Explain why this has happened.

10) What is rotational slip?

11) Explain what freeze-thaw weathering is.

12) What is a corrie?

13) How does a pyramidal peak form?

14) Give an example of a pyramidal peak.

15) Explain how a hanging valley forms.

16) What is bulldozing?

17) Give one difference between lateral and ground moraine.

18) Where is medial moraine deposited?

19) Describe what a drumlin looks like.

20) How would you identify a pyramidal peak on a map?

21) Describe what a glacial trough looks like on a map.

22) Give an example of a glacial trough.

23) What does a ribbon lake look like on a map?

24) Give one reason why areas covered in snow and ice are fragile environments.

25) Give two social impacts of tourism on areas covered in snow and ice.

26) Describe two environmental impacts of tourism on areas covered in snow and ice.

27) a) Give an example of an area in the Alps used for winter sports and sightseeing of glaciers.

   b) Give one economic, one social and one environmental impact of tourism in the area you named.

   c) Give three management strategies used in the area you named.

28) Name an industry affected by glacial retreat.

29) Give one economic impact of glacial retreat.

30) Give one social impact of glacial retreat.

Unit 1B — The Coastal Zone

Coastal Weathering and Erosion

Weathering is the breakdown of rocks where they are, erosion is when the rocks are broken down and carried away by something, e.g. by seawater. Poor coastal zone, I bet it's worn down.

Rock is Broken Down by Mechanical and Chemical Weathering

1) Mechanical weathering is the breakdown of rock without changing its chemical composition. There's one main type of mechanical weathering that affects coasts — freeze-thaw weathering:

> 1) It happens when the temperature alternates above and below 0 °C (the freezing point of water).
> 2) Water gets into rock that has cracks, e.g. granite.
> 3) When the water freezes it expands, which puts pressure on the rock.
> 4) When the water thaws it contracts, which releases the pressure on the rock.
> 5) Repeated freezing and thawing widens the cracks and causes the rock to break up.

2) Chemical weathering is the breakdown of rock by changing its chemical composition. Carbonation weathering is a type of chemical weathering that happens in warm and wet conditions:

> 1) Rainwater has carbon dioxide dissolved in it, which makes it a weak carbonic acid.
> 2) Carbonic acid reacts with rock that contains calcium carbonate, e.g. carboniferous limestone, so the rocks are dissolved by the rainwater.

Mass movement is when Material Shifts Down a Slope as One

1) Mass movement is the shifting of rocks and loose material down a slope, e.g. a cliff. It happens when the force of gravity acting on a slope is greater than the force supporting it.

2) Mass movements cause coasts to retreat rapidly.

3) They're more likely to happen when the material is full of water — it acts as a lubricant.

4) You need to know about two types of mass movement.

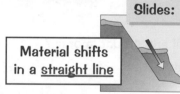

Slides:
Material shifts in a straight line

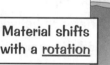

Slumps:
Material shifts with a rotation

Waves Wear Away the Coast Using Four Processes of Erosion

> 1) Hydraulic power — waves crash against rock and compress the air in the cracks. This puts pressure on the rock. Repeated compression widens the cracks and makes bits of rock break off.
> 2) Abrasion — eroded particles in the water scrape and rub against rock, removing small pieces.
> 3) Attrition — eroded particles in the water smash into each other and break into smaller fragments. Their edges also get rounded off as they rub together.
> 4) Solution — weak carbonic acid in seawater dissolves rock like chalk and limestone.

The waves that carry out erosional processes are called destructive waves:

1) Destructive waves have a high frequency (10-14 waves per minute).

2) They're high and steep.

3) Their backwash (the movement of the water back down the beach) is more powerful than their swash (the movement of the water up the beach). This means material is removed from the coast.

High, steep wave

Backwash   Swash

If you find yourself slumping — have a little break from revision...

This page is packed full of information, but it's really only about how the coast is worn away and rocks are broken down into smaller pieces. Break your revision down into smaller pieces by learning the processes one at a time.

Coastal Landforms Caused by Erosion

Erosion by waves forms many <u>coastal landforms</u> over <u>long periods of time</u>.

Waves Erode Cliffs to Form Wave-cut Platforms

1) Waves cause <u>most erosion</u> at the <u>foot</u> of a cliff (see diagrams below).
2) This forms a <u>wave-cut notch</u>, which is enlarged as <u>erosion</u> continues.
3) The rock above the notch becomes <u>unstable</u> and eventually <u>collapses</u>.
4) The <u>collapsed material</u> is washed away and a <u>new</u> wave-cut notch starts to form.
5) <u>Repeated collapsing</u> results in the <u>cliff retreating</u>.
6) A <u>wave-cut platform</u> is the platform that's <u>left behind</u> as the <u>cliff retreats</u>.

There are cliffs and wave-cut platforms at Beachy Head in Sussex.

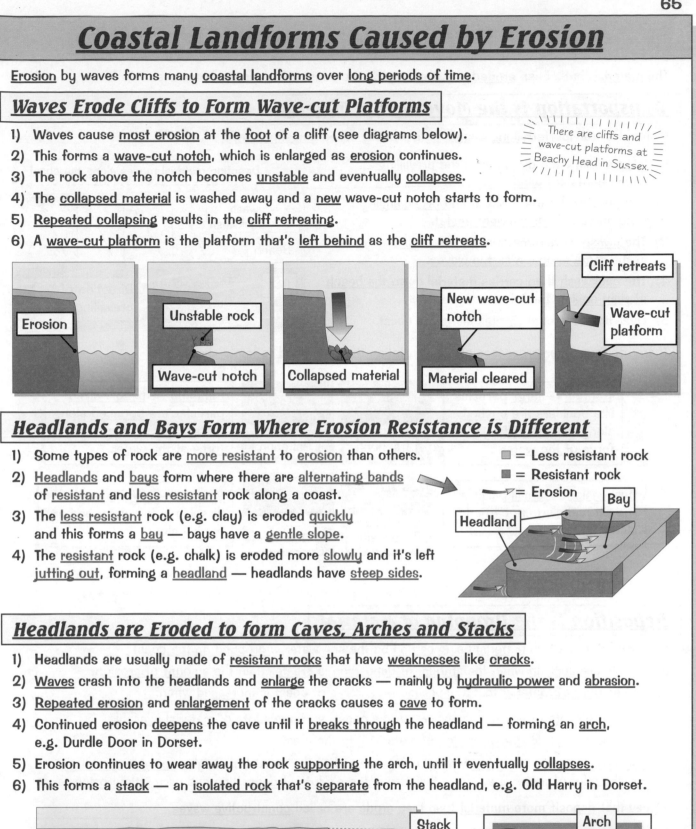

Headlands and Bays Form Where Erosion Resistance is Different

1) Some types of rock are <u>more resistant</u> to <u>erosion</u> than others.
2) <u>Headlands</u> and <u>bays</u> form where there are <u>alternating bands</u> of <u>resistant</u> and <u>less resistant</u> rock along a coast.
3) The <u>less resistant</u> rock (e.g. clay) is eroded <u>quickly</u> and this forms a <u>bay</u> — bays have a <u>gentle slope</u>.
4) The <u>resistant</u> rock (e.g. chalk) is eroded more <u>slowly</u> and it's left <u>jutting out</u>, forming a <u>headland</u> — headlands have <u>steep sides</u>.

■ = Less resistant rock
■ = Resistant rock
⇗ = Erosion

Bay
Headland

Headlands are Eroded to form Caves, Arches and Stacks

1) Headlands are usually made of <u>resistant rocks</u> that have <u>weaknesses</u> like <u>cracks</u>.
2) <u>Waves</u> crash into the headlands and <u>enlarge</u> the cracks — mainly by <u>hydraulic power</u> and <u>abrasion</u>.
3) <u>Repeated erosion</u> and <u>enlargement</u> of the cracks causes a <u>cave</u> to form.
4) Continued erosion <u>deepens</u> the cave until it <u>breaks through</u> the headland — forming an <u>arch</u>, e.g. Durdle Door in Dorset.
5) Erosion continues to wear away the rock <u>supporting</u> the arch, until it eventually <u>collapses</u>.
6) This forms a <u>stack</u> — an <u>isolated rock</u> that's <u>separate</u> from the headland, e.g. Old Harry in Dorset.

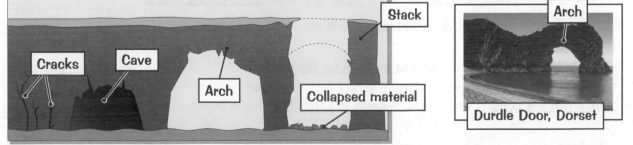

Stack
Cracks
Cave
Arch
Collapsed material
Arch
Durdle Door, Dorset

What do you call a man with a seagull on his head? Cliff...

<u>Crack erosion</u> — sounds painful. If you don't want it to happen to you, then don't stand in the sea for too many years in a row. Just a few <u>landforms</u> to learn here, and as ever, learning the diagrams will help in the exam.

Coastal Transportation and Deposition

The <u>material</u> that's been <u>eroded</u> is <u>moved around</u> the coast and <u>deposited</u> by waves.

Transportation is the Movement of Material

Material is transported <u>along coasts</u> by a process called <u>longshore drift</u>:

1) <u>Waves</u> follow the <u>direction</u> of the <u>prevailing</u> (most common) <u>wind</u>.

2) They usually hit the coast at an <u>oblique angle</u> (any angle that <u>isn't a right angle</u>).

3) The <u>swash</u> carries material <u>up the beach</u>, in the <u>same direction as the waves</u>.

4) The <u>backwash</u> then carries material <u>down the beach</u> at <u>right angles</u>, back towards the sea.

5) Over time, material <u>zigzags</u> along the coast.

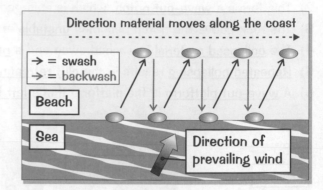

There are <u>four</u> other <u>processes of transportation</u> that you need to know about:

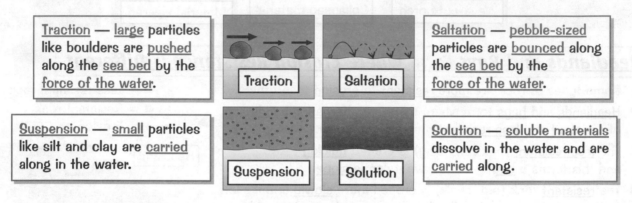

<u>Traction</u> — <u>large</u> particles like boulders are <u>pushed</u> along the <u>sea bed</u> by the <u>force of the water</u>.

Traction | Saltation

<u>Saltation</u> — <u>pebble-sized</u> particles are <u>bounced</u> along the <u>sea bed</u> by the <u>force of the water</u>.

<u>Suspension</u> — <u>small</u> particles like silt and clay are <u>carried</u> along in the water.

Suspension | Solution

<u>Solution</u> — <u>soluble materials</u> dissolve in the water and are <u>carried</u> along.

Deposition is the Dropping of Material

1) Deposition is when <u>material</u> being <u>carried</u> by the sea water is <u>dropped on the coast</u>.

2) Coasts are <u>built up</u> when the <u>amount of deposition</u> is <u>greater</u> than the <u>amount of erosion</u>.

3) The <u>amount of material</u> that's <u>deposited</u> on an area of coast is <u>increased</u> when:

- There's <u>lots</u> of <u>erosion</u> elsewhere on the coast, so there's <u>lots of material available</u>.
- There's <u>lots</u> of <u>transportation</u> of material <u>into</u> the area.

4) <u>Low energy</u> waves (i.e. <u>slow</u> waves) carry material to the coast but they're <u>not strong enough</u> to take a lot of material away — this means there's <u>lots of deposition</u> and <u>very little erosion</u>.

Waves that <u>deposit more material</u> than they <u>erode</u> are called <u>constructive waves</u>.

1) Constructive waves have a <u>low frequency</u> (6-8 waves per minute).

2) They're <u>low</u> and <u>long</u>.

3) The <u>swash</u> is <u>powerful</u> and it <u>carries material up the coast</u>.

4) The backwash is <u>weaker</u> and it <u>doesn't</u> take a lot of material <u>back down the coast</u>. This means material is <u>deposited</u> on the coast.

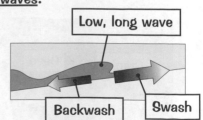

<u>Why did the constructive wave go to the bank? It wanted to make a deposit...</u>

Some more processes to learn here but none of them are tricky. You might find it useful to draw yourself a <u>diagram</u> of how <u>longshore drift</u> works — you'll get a feel for how the material is <u>moved along</u> the coast in a <u>zigzag</u> pattern.

Coastal Landforms Caused by Deposition

Here are some more exciting <u>landforms</u> for you to learn about. This time it's all about <u>deposition</u>. Unfortunately you're going to be slightly disappointed — sandcastles won't be in the exam.

Beaches are formed by Deposition

1) Beaches are found on coasts <u>between</u> the <u>high water mark</u> (the <u>highest point on the land</u> the <u>sea level</u> gets to) and the <u>low water mark</u> (the <u>lowest point</u> on the land the <u>sea level</u> gets to).

2) They're formed by <u>constructive waves</u> (see p. 66) depositing material like <u>sand</u> and <u>shingle</u>.

3) <u>Sand</u> and <u>shingle beaches</u> have different <u>characteristics</u>:

High water mark · Constructive wave
Low water mark · Beach

- <u>Sand</u> beaches are <u>flat</u> and <u>wide</u> — sand particles are <u>small</u> and the weak backwash <u>can</u> move them <u>back down</u> the beach, creating a <u>long</u>, <u>gentle slope</u>.
- <u>Shingle</u> beaches are <u>steep</u> and <u>narrow</u> — shingle particles are <u>large</u> and the weak backwash <u>can't</u> move them back down the beach. The shingle particles <u>build up</u> and create a <u>steep slope</u>.

Spits and Bars are formed by Longshore Drift

Spits are just <u>beaches</u> that <u>stick out</u> into the sea — they're <u>joined</u> to the coast at <u>one end</u>. If a spit sticks out so far that it <u>connects</u> with another bit of the mainland, it'll form a <u>bar</u>. Spits and bars are formed by the process of <u>longshore drift</u> (see p. 66).

SPITS

1) Spits form at <u>sharp bends</u> in the coastline, e.g. at a <u>river mouth</u>.

2) <u>Longshore drift</u> transports sand and shingle <u>past</u> the bend and <u>deposits</u> it in the sea.

3) Strong winds and waves can <u>curve</u> the end of the spit (forming a <u>recurved end</u>).

4) The <u>sheltered area</u> behind the spit is <u>protected from waves</u> — lots of material <u>accumulates</u> in this area, which means <u>plants</u> can grow there.

5) <u>Over time</u>, the sheltered area can become a <u>mud flat</u> or a <u>salt marsh</u>.

An example of a spit is Spurn Head in Yorkshire.

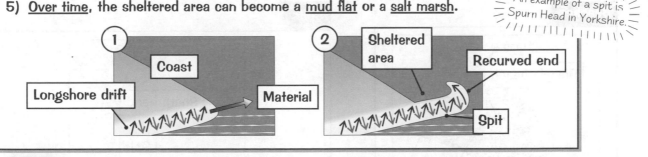

1 · Coast · Longshore drift · Material
2 · Sheltered area · Recurved end · Spit

BARS

1) A bar is formed when a spit <u>joins two headlands together</u>, e.g. there's a bar at Slapton in Devon.

2) The bar <u>cuts off</u> the bay between the headlands <u>from the sea</u>.

3) This means a <u>lagoon</u> can form <u>behind</u> the bar.

4) There's not much else to say about them.

1 · Bay · Spit
2 · Lagoon · Bar

Depositional bars — the only cocktail they serve is a long beach iced tea...

The things you learn in geography are life skills. If you have to meet someone at the <u>beach</u> you now know exactly where to stand — between the <u>high</u> and <u>low water marks</u>, of course. And who said geography was just about maps...

Coastal Landforms on Maps

I love <u>maps</u>, all geographers love maps. I can't get to sleep unless I've got one under my pillow.
So I'm going to do you a favour and share my passion with you — check out these <u>coastal landforms</u>...

Identifying Landforms Caused by Erosion

You might be asked to <u>identify coastal landforms</u> on a <u>map</u> in the exam. The simplest
thing they could ask is whether the map is showing <u>erosional</u> or <u>depositional landforms</u>,
so here's how to <u>identify</u> a few <u>erosional landforms</u> to get you started:

Have a gander at pages 144-145 for more on reading maps.

Caves, arches and stacks

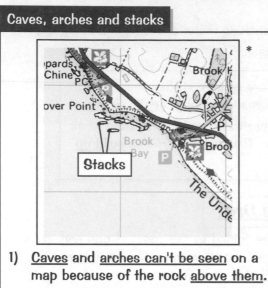

1) <u>Caves</u> and <u>arches can't be seen</u> on a map because of the rock <u>above them</u>.
2) <u>Stacks</u> look like <u>little blobs</u> in the sea.

Cliffs and wave-cut platforms

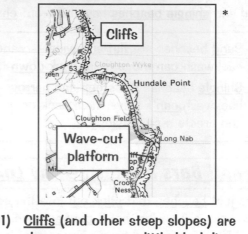

1) <u>Cliffs</u> (and other steep slopes) are shown on maps as <u>little black lines</u>.
2) <u>Wave-cut platforms</u> are shown as <u>bumpy edges</u> along the coast.

Identifying Landforms Caused by Deposition

<u>Identifying depositional landforms</u> is easy once you know that <u>beaches</u> are shown in <u>yellow</u> on maps.
Here's how to <u>identify</u> a couple of <u>depositional landforms</u>:

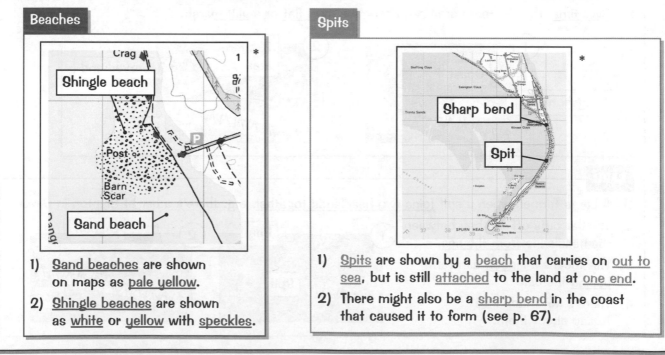

Beaches

1) <u>Sand beaches</u> are shown on maps as <u>pale yellow</u>.
2) <u>Shingle beaches</u> are shown as <u>white</u> or <u>yellow</u> with <u>speckles</u>.

Spits

1) <u>Spits</u> are shown by a <u>beach</u> that carries on <u>out to sea</u>, but is still <u>attached</u> to the land at <u>one end</u>.
2) There might also be a <u>sharp bend</u> in the coast that caused it to form (see p. 67).

Find the spit on the map — and then wipe it off...

There are some seriously easy marks up for grabs with map questions so make sure you learn this page. You could
practise looking for <u>landforms</u> on any <u>maps</u> you can get a hold of. Probably best to avoid the inner-city maps though.

Unit 1B — The Coastal Zone

Rising Sea Level and Coastal Flooding

Rising sea level — there's no need to worry if you live in a block of flats... or on top of Mount Everest.

Sea Level is Rising because of Global Warming

Global sea level is rising at a rate of about 2 mm per year. That might not sound like a lot, but sea level has increased by about 20 cm over the past century. It's predicted to rise by between 18 and 59 cm by 2100. The cause of rising sea level is global warming — the rapid rise in global temperature over the last 100 years. Global warming has two effects that cause sea level to rise:

1) **Melting ice**

   The melting of ice on land (e.g. the Antarctic ice sheet) causes water that's stored as ice to return to the oceans. This increases the volume of water in the oceans and causes sea level to rise.

2) **Heating oceans**

   Increased global temperature causes the oceans to get warmer and expand (thermal expansion). This increases the volume of water, causing sea level to rise.

Rising Sea Level will Increase Coastal Flooding

Rising sea level will mean coastal flooding will happen more often and will cause more damage, especially in low-lying parts of the world like Bangladesh and the Maldives. Coastal flooding has a variety of impacts:

Economic

1) Loss of tourism — many coastal areas are popular tourist destinations. Flooding can cause tourist attractions to close and it can put people off visiting.
2) Damage repair — repairing flood damage can be extremely expensive.
3) Loss of agricultural land — seawater has a high salt content. Salt reduces soil fertility, so crop production can be affected for years after a flood.

Social

1) Deaths — coastal floods have killed thousands of people in the past.
2) Water supplies affected — floodwater can pollute drinking water with salt or sewage.
3) Loss of housing — many people are made homeless because of floods.
4) Loss of jobs — coastal industries may be shut down because of damage to equipment and buildings, e.g. fishing boats can be destroyed.

Environmental

1) Ecosystems affected — seawater has a high salt content. Increased salt levels can damage or kill organisms in an ecosystem.
2) Vegetation killed by water — the force of floodwater uproots trees and plants. Standing flood water also drowns some trees and plants.
3) Increased erosion — a large volume of fast-moving water can erode lots of material, damaging the environment.

Political

The government has to make policies to reduce the impacts of future flooding. They can do things like building more or better flood defences, or they can manage the use of areas that might be flooded, e.g. by stopping people living there.

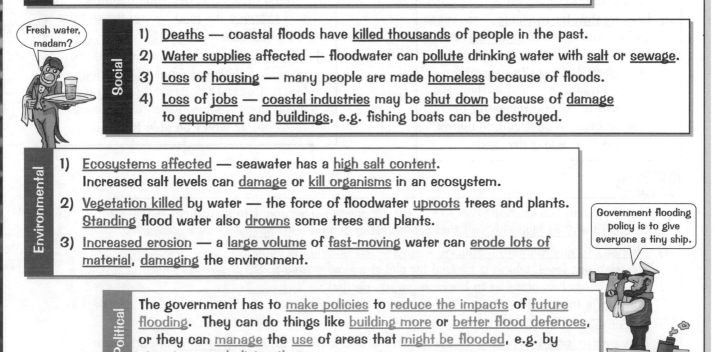

Fresh water, madam?

Government flooding policy is to give everyone a tiny ship.

The Antarctic ice sheet is the biggest ice cube in the world...

As you've just discovered, coastal flooding can have enormous impacts on coastal areas. But I know what you're thinking, something's missing. I know what it is — a case study on coastal flooding and it's only a page away.

Coastal Flooding — Case Study

Another case to study — this one's all about the impacts of coastal flooding on the Maldives. Enjoy...

The Maldives is a Group of Islands in the Indian Ocean

Population: About 300 000 people.

Number of islands: 1190, of which 199 are inhabited.

Average island height: 1.5 m above sea level — 80% of the land is below 1 m. Because of rising sea levels, scientists think the islands will be completely submerged within 50 to 100 years.

Coastal Flooding has a Variety of Impacts on the Maldives

Economic

1) Loss of tourism — tourism is the largest industry in the Maldives. If the main airport can't work properly because of coastal flooding the country will be cut off from international tourists. This will massively reduce the country's income.

2) Disrupted fishing industry — fish are the Maldives' largest export. Coastal flooding may damage fish processing plants, reducing the fish exports and the country's income.

Social

1) Houses damaged or destroyed — a severe flood could make entire communities homeless.

2) Less freshwater available — supplies of freshwater are already low on many of the islands. If supplies are polluted with salty seawater during floods, then some islands will have to rely on rainwater or build expensive desalination plants to meet their water demands.

Environmental

1) Loss of beaches — coastal flooding wears away beaches on the islands at a rapid rate. This destroys habitats and exposes the land behind the beach to the effects of flooding.

2) Loss of soil — the soil on most of the islands is shallow (about 20 cm deep or less). Coastal floods could easily wash away the soil layer, which would mean most plants won't be able to grow.

Political

1) The Maldivian Government had to ask the Japanese Government to give them $60 million to build the 3 m high sea wall that protects the capital city, Malé.

2) Changes to environmental policies — increased flooding is caused by rising sea level, which is caused by global warming (see p. 69). The Maldives has pledged to become carbon neutral so it doesn't contribute to global warming. The Maldivian Government is encouraging other governments to do the same.

Carbon neutral means not adding carbon dioxide (CO_2) to the atmosphere — CO_2 is thought to be causing global warming.

3) Changes to long-term plans — the government is thinking about buying land in countries like India and Australia and moving Maldivians there, before the islands become uninhabitable.

It's a lovely place to visit — if you've got a submarine...

The Maldives are as flat as a pancake and that's not ideal when you're surrounded by the sea. Now you've read this page you'll know about the impacts that coastal flooding is having on the country. In short, it's not looking good...

Coastal Erosion — Case Study

Holderness in east Yorkshire has one of the fastest eroding coastlines in Europe. What a claim to fame...

The Average Rate of Erosion at Holderness is About 1.8 Metres per Year

1) The Holderness coastline is 61 km long — it stretches from Flamborough Head (a headland) to Spurn Head (a spit).

2) Erosion is causing the cliffs to collapse along the coastline. The material then gets washed away, so the coastline is retreating.

3) About 1.8 m of land is lost to the sea every year — in some places, e.g. Great Cowden, the rate of erosion has been over 10 m per year in recent years.

4) You need to know the main reasons for this rapid erosion at Holderness and the impact it has on people's lives and the environment:

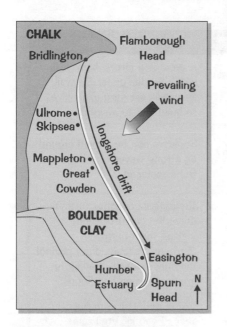

Main reasons for rapid erosion

1) Easily eroded rock type — the cliffs are mostly made of boulder clay which is easily eroded. It's likely to slump (see p. 64) when it's wet, causing the cliffs to collapse.

2) Naturally narrow beaches — beaches slow waves down, reducing their erosive power so narrow beaches give less protection.

3) People worsening the situation — coastal defences called groynes (see p. 72) have been built at Mappleton. Groynes stop material from being moved further down the coast. This means the beaches are narrower and more easily eroded in some other places.

4) Powerful waves — Holderness faces the prevailing wind direction, which brings waves from the north east (all the way from the Arctic Ocean). Waves increase in power over this long distance, so the coast is battered by highly erosive waves.

Impacts on people's lives

1) Homes near the cliffs (e.g. in Skipsea) are at risk of collapsing into the sea.

2) Property prices along the coast have fallen sharply for those houses at risk from erosion.

3) Accessibility to some settlements has been affected because roads near the cliff tops are at risk of collapsing into the sea, e.g. Southfield Lane which runs between Skipsea and Ulrome has been closed.

4) Businesses are at risk from erosion so people will lose their jobs, e.g. Seaside Caravan Park at Ulrome is losing an average of 10 pitches a year.

5) The gas terminal at Easington is at risk (it's only 25 m from the cliff edge). This terminal accounts for 25% of Britain's gas supply.

6) 80 000 m² of farmland is lost each year. This has a huge effect on farmers' livelihoods.

Environmental impacts

Some SSSIs (Sites of Special Scientific Interest) are threatened — e.g. the Lagoons near Easington are part of an SSSI. The Lagoons are separated from the sea by a narrow strip of sand and shingle (a bar). If this is eroded it will connect the Lagoons to the sea and they would be destroyed.

Extreme caravanning — the latest craze in Ulrome...

Blimey, Holderness really is taking a battering from the sea. Make sure you can remember the causes and the impacts of the rapid erosion — it's unlikely that you'll only be asked about one of them in an exam. To the next page...

Coastal Management Strategies

The <u>aim</u> of coastal management is to <u>protect</u> people and the environment from the <u>impacts</u> of erosion and flooding. <u>Not all</u> coastal areas can be managed though — the amount of <u>money</u> available is <u>limited</u>.

Coastal Defences Include Hard and Soft Engineering

There are <u>two</u> types of strategy to <u>deal with coastal flooding</u> and <u>erosion</u>:

<u>Hard engineering</u> — <u>man-made structures</u> built to <u>control the flow</u> of the sea and <u>reduce flooding</u> and <u>erosion</u>.

<u>Soft engineering</u> — schemes set up using <u>knowledge</u> of the sea and its <u>processes</u> to <u>reduce the effects of flooding</u> and <u>erosion</u>.

There's <u>debate</u> about <u>which strategies are best</u>, so you'll need to know the <u>benefits</u> and <u>disadvantages</u> of a few of them. Here's a lovely big table for you to enjoy:

	Strategy	What it is	Benefits	Disadvantages
HARD ENGINEERING	Sea wall	A <u>wall</u> made out of a <u>hard material</u> like <u>concrete</u> that <u>reflects</u> waves back to sea.	It <u>prevents erosion</u> of the coast. It also acts as a <u>barrier</u> to <u>prevent flooding</u>.	It creates a <u>strong backwash</u>, which <u>erodes under</u> the wall. Sea walls are <u>very expensive</u> to <u>build</u> and to <u>maintain</u>.
	Rock armour	<u>Boulders</u> that are <u>piled up</u> along the coast.	The boulders <u>absorb wave energy</u> and so <u>reduce erosion</u> and <u>flooding</u>. It's a fairly <u>cheap</u> defence.	Boulders can be <u>moved around</u> by <u>strong waves</u>, so they need to be <u>replaced</u>.
	Groynes	Wooden or stone <u>fences</u> that are built at <u>right angles</u> to the coast. They <u>trap material</u> transported by <u>longshore drift</u>.	Groynes create <u>wider beaches</u> which <u>slow</u> the <u>waves</u>. This gives greater <u>protection</u> from <u>flooding</u> and <u>erosion</u>. They're a fairly <u>cheap</u> defence.	They <u>starve beaches</u> further down the coast of sand, making them <u>narrower</u>. Narrower beaches <u>don't protect</u> the coast as well, leading to <u>greater erosion</u> and <u>floods</u>.
SOFT ENGINEERING	Beach nourishment	Sand and shingle from <u>elsewhere</u> (e.g. the <u>offshore seabed</u>) that's <u>added</u> to beaches.	Beach nourishment creates <u>wider beaches</u> which <u>slow</u> the <u>waves</u>. This gives greater <u>protection</u> from <u>flooding</u> and <u>erosion</u>.	Taking <u>material</u> from the <u>seabed</u> can <u>kill</u> organisms like <u>sponges</u> and <u>corals</u>. It's a <u>very expensive</u> defence. It has to be <u>repeated</u>.
	Dune regeneration	<u>Creating</u> or <u>restoring sand dunes</u> by either <u>nourishment</u>, or <u>by planting vegetation</u> to <u>stabilise</u> the sand.	Sand dunes provide a <u>barrier</u> between the land and the sea. <u>Wave energy</u> is <u>absorbed</u> which <u>prevents flooding</u> and <u>erosion</u>. <u>Stabilisation</u> is <u>cheap</u>.	The <u>protection</u> is <u>limited</u> to a <u>small area</u>. <u>Nourishment</u> is <u>very expensive</u>.
	Marsh creation	Planting <u>vegetation</u> in <u>mudflats</u> along the coast.	The vegetation <u>stabilises</u> the mudflats and helps to <u>reduce</u> the <u>speed</u> of the waves. This <u>prevents flooding</u> and <u>erosion</u>. It also creates <u>new habitats</u> for organisms.	Marsh creation <u>isn't useful</u> where <u>erosion rates</u> are <u>high</u> because the marsh can't <u>establish itself</u>. It's a fairly <u>expensive</u> defence.
	Managed retreat	<u>Removing</u> an <u>existing defence</u> and allowing the land behind it to flood.	<u>Over time</u> the land will become <u>marshland</u> — creating <u>new habitats</u>. <u>Flooding</u> and <u>erosion</u> are <u>reduced</u> behind the marshland. It's a fairly <u>cheap</u> defence.	People may <u>disagree</u> over what land is <u>allowed to flood</u>, e.g. flooding farmland would affect the <u>livelihood</u> of farmers.

<u>A good management strategy involves warm-ups — they reduce groyne strains...</u>

Wow, that sure is a mighty fine table. It seems like a lot to remember but I promise you, it's really not that tough. Make sure you know at least a couple of <u>benefits</u> and <u>disadvantages</u> for <u>each strategy</u> so you'll be a winner in the exam.

Coastal Management — Case Study

We're going back to <u>Holderness</u> to find out what <u>management strategies</u> are being used there. You know, it's possible that we're having too much fun...

Hard Engineering Strategies have been used Along Holderness

Page 71 outlines the <u>main reasons</u> for the rapid erosion along Holderness and the impacts it's having. To try to <u>reduce</u> the <u>effects</u> of erosion, <u>11.4 km</u> of Holderness coastline has been <u>protected</u> by <u>hard engineering</u>:

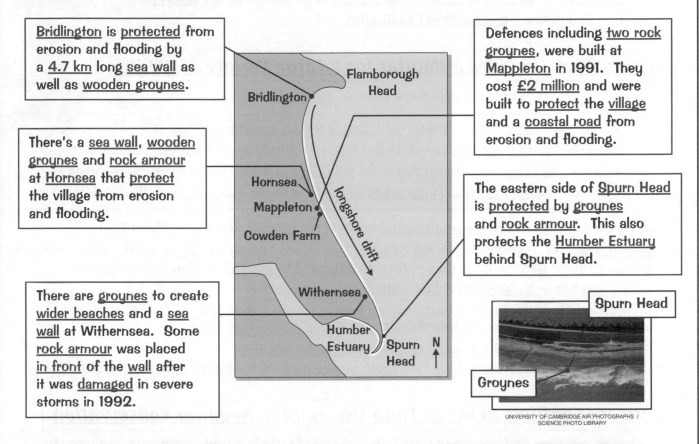

<u>Bridlington</u> is <u>protected</u> from erosion and flooding by a <u>4.7 km</u> long <u>sea wall</u> as well as <u>wooden groynes</u>.

Defences including <u>two rock groynes</u>, were built at <u>Mappleton</u> in 1991. They cost <u>£2 million</u> and were built to <u>protect</u> the <u>village</u> and a <u>coastal road</u> from erosion and flooding.

There's a <u>sea wall</u>, <u>wooden groynes</u> and <u>rock armour</u> at <u>Hornsea</u> that <u>protect</u> the village from erosion and flooding.

The eastern side of <u>Spurn Head</u> is <u>protected</u> by <u>groynes</u> and <u>rock armour</u>. This also protects the <u>Humber Estuary</u> behind Spurn Head.

There are <u>groynes</u> to create <u>wider beaches</u> and a <u>sea wall</u> at Withernsea. Some <u>rock armour</u> was placed <u>in front</u> of the <u>wall</u> after it was <u>damaged</u> in severe storms in 1992.

Bridlington · Flamborough Head · Hornsea · Mappleton · Cowden Farm · longshore drift · Withernsea · Humber Estuary · Spurn Head · N

Spurn Head

Groynes

UNIVERSITY OF CAMBRIDGE AIR PHOTOGRAPHS / SCIENCE PHOTO LIBRARY

The Strategies are Locally Successful but Cause Problems Elsewhere

1) <u>Groynes</u> protect <u>local areas</u> but cause <u>narrow beaches</u> to form <u>further down the Holderness coast</u>. This increases erosion down the coast, e.g. <u>Cowden Farm</u> (<u>south</u> of <u>Mappleton</u>) is now <u>at risk</u> of <u>falling into the sea</u>.

2) The <u>material produced</u> from the erosion of Holderness is <u>normally transported south</u> into the <u>Humber Estuary</u> and <u>down</u> the <u>Lincolnshire coast</u>. Reducing the amount of material that's eroded and transported south <u>increases</u> the <u>risk</u> of <u>flooding</u> in the Humber Estuary, because there's <u>less material</u> to slow the floodwater down.

3) The rate of <u>coastal retreat</u> along the <u>Lincolnshire coast</u> is also increased, because <u>less new material</u> is being added.

4) <u>Spurn Head</u> is <u>at risk</u> of <u>being eroded away</u> because <u>less material</u> is being <u>added to it</u>.

5) <u>Bays</u> are forming <u>between</u> the <u>protected areas</u>, and the protected areas are becoming <u>headlands</u> which are being <u>eroded more heavily</u>. This means <u>maintaining the defences</u> in the protected areas is becoming <u>more expensive</u>.

Holderness coastal management officer — probably not the easiest job to have...

This follows on from the <u>case study</u> on p. 71, so hopefully most of the place names seem familiar. You'll probably be able to use <u>information</u> from <u>both pages</u> to answer a question on <u>coastal management</u>. What an educational bargain.

Coastal Habitat — Case Study

Coastal areas get pretty heavily used by people (beach bums), but they're important for wildlife too...

Studland Bay is a Coastal Area with Beaches, Dunes and Heathland

1) Studland Bay is a bay in Dorset, in the south west of England.
2) It's mostly sheltered from highly erosive waves, but the southern end of the bay is being eroded.
3) There are sandy beaches around the bay, with sand dunes and heathland behind them.
4) The heathland is a Site of Special Scientific Interest (SSSI) and a nature reserve.
5) Studland Bay is also a popular tourist destination.

Studland Bay Provides a Habitat for a Large Variety of Wildlife

Here are a few examples of the wildlife that's found in Studland Bay:

- Reptiles like adders, grass snakes, sand lizards and slow worms.
- Birds like Dartford warblers (a rare bird in England), shelducks and grebes.
- Fish like seahorses — Studland Bay is the only place in Britain where the spiny seahorse breeds.
- Plants like marram grass and lyme grass on the sand dunes and heather on the heathland.

Some of these organisms are specially adapted to live in the habitats found in Studland Bay:

1) Marram grass has folded leaves to reduce water loss — sand dunes are windy and dry which increases transpiration. It also has long roots to take up water and to stabilise itself in the loose sand.
2) Lyme grass has waxy leaves to reduce water loss by transpiration.
3) Grebes — these birds dive underwater to find food in the sea. Their feet are far back on their bodies to help them dive (it makes them streamlined).
4) Snakes and lizards have thick, scaly skin to reduce water loss from their bodies. It also protects them from rough undergrowth on the heathland.

Transpiration is the loss of water from plants by evaporation.

There are Conflicts Between Land Use and the Need for Conservation

Some human activities (e.g. recreation) don't use the environment in a sustainable way (they use up resources or damage the environment). The environment is managed to make sure it's conserved, but can also be used for other activities:

1) Lots of people walk across the sand dunes which has caused lots of erosion. The National Trust manages the area so people can use the sand dunes without damaging them too much:

- Boardwalks are used to guide people over the dunes so the sand beneath them is protected.
- Some sand dunes have been fenced off and marram grass has been planted in them. This gives the dunes a chance to recover and the marram grass stabilises the sand.
- Information signs have been put up to let visitors know why the sand dune habitat is important, and how they can enjoy the environment without damaging it.

2) Hundreds of boats use Studland Bay and their anchors are destroying the seagrass where seahorses live. Seahorses are protected by law, so boat owners are being told to not damage the seagrass.
3) The heathland behind the sand dunes is an important habitat, but it can be damaged by fires caused by things like cigarettes, e.g. in 2008 a fire destroyed six acres of heathland. The National Trust is educating visitors on the dangers of causing fires and has provided fire beaters to extinguish flames.

Stud-land bay — sorry girls, it's not what you're thinking of...

Plenty of words here and not very many pretty pictures I'm afraid. It'll just be a case of cramming all these facts into your head so you can cough them up in the exam. Just think of the seahorses and it'll all be worth it in the end.

Revision Summary for The Coastal Zone

So, you've coasted through another topic — that means it's time to find out just how much of this information has been deposited in your noggin. Have a go at the questions below. If you're finding it tough, just look back at the pages in the topic and then have another go. You'll be ready to move on when you can answer all of these questions without breaking sweat.

1) What is mechanical weathering?

2) Describe the process of chemical weathering.

3) a) What is mass movement?

   b) Give an example of one type of mass movement.

4) How do waves erode the coast by hydraulic power?

5) What waves are associated with coastal erosion?

6) Describe how a wave-cut platform is formed.

7) Are headlands made of more or less resistant rock?

8) Describe how erosion can turn a crack in a cliff into a cave.

9) By what process is material transported along coasts?

10) What is deposition?

11) What waves are associated with coastal deposition?

12) Where is a beach formed on a coast?

13) Why is a sand beach flatter and wider than a shingle beach?

14) Name the transportation process that forms spits and bars.

15) Where do spits form?

16) Why can't cracks, caves and arches be seen on a map?

17) How are cliffs shown on a map?

18) On maps, what do speckles on top of yellow shading tell you?

19) Describe one effect of global warming that's causing sea level to rise.

20) Give two economic impacts of coastal flooding.

21) Give two social impacts of coastal flooding.

22) Why is the risk of coastal flooding becoming greater in the Maldives?

23) Describe a political impact of coastal flooding in the Maldives.

24) a) Give two main reasons for rapid erosion along a named coastline.

   b) Describe three impacts on people's lives that are caused by erosion of the same coastline.

   c) Describe an environmental impact caused by erosion of the same coastline.

25) Describe the difference between hard engineering and soft engineering coastal management strategies.

26) Explain a disadvantage of using groynes as a coastal defence.

27) a) Name two soft engineering strategies.

   b) Give one benefit of each strategy.

28) a) Give two examples of hard engineering strategies used along a named coastline.

   b) Describe two problems caused by the use of hard engineering strategies along the same coastline.

29) a) Describe how two organisms are adapted to living in a named coastal habitat.

   b) Describe two strategies for dealing with conflicts between land use and conservation in this coastal habitat.

Population Growth

The <u>population change</u> topic — putting the 'pop' in 'popular'. Studying population change is amazing — you'd have to be stark raving mad to not agree with me. Only one way to judge for yourself though...

The World's Population is Growing Rapidly

1) The graph shows <u>world population</u> for the years <u>1500-2000</u> — it's <u>been increasing</u> and is <u>still increasing today</u>.

2) The population of the world is <u>increasing</u> at an <u>exponential rate</u> — it's growing <u>faster and faster</u>.

3) There are <u>two things</u> that affect the <u>population size</u> of the world:

> <u>Birth rate</u> — the number of live babies born per thousand of the population per year.
>
> <u>Death rate</u> — the number of deaths per thousand of the population per year.

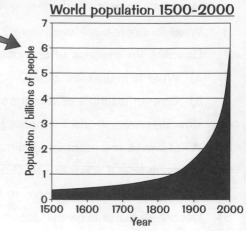
World population 1500-2000

4) When the <u>birth rate</u> is <u>higher</u> than the death rate, more people are being <u>born</u> than are <u>dying</u>, so the <u>population grows</u> — this is called the <u>natural increase</u>.

5) It's called the <u>natural decrease</u> when the <u>death rate's higher</u> than the birth rate.

6) The population size of a <u>country</u> is also affected by <u>migration</u> — the <u>movement</u> of <u>people</u> from <u>one area</u> to <u>another area</u>.

Countries go Through Five Stages of Population Growth

Countries go through <u>five different stages</u> of <u>population growth</u>.
These are shown by the <u>Demographic Transition Model</u> (<u>DTM</u>):

	Stage 1	Stage 2	Stage 3	Stage 4	Stage 5
Birth rate	High and fluctuating	High and steady	Rapidly falling	Low and fluctuating	Slowly falling
Death rate	High and fluctuating	Rapidly falling	Slowly falling	Low and fluctuating	Low and fluctuating
Population growth rate	Zero	Very high	High	Zero	Negative
Population size	Low and steady	Rapidly increasing	Increasing	High and steady	Slowly falling

<u>Poorer</u>, <u>less developed</u> countries are in the <u>earlier stages</u> of the DTM, whilst <u>richer</u>, <u>more developed</u> countries are in the <u>later stages</u>.

You know what — this world population is growing on me...

<u>DTM</u> — it stands for Done Too Much (work that is). Sigh, it's a pretty useful thing to know about when you're studying <u>population change</u> though, and it's not too tough to learn, so <u>copy it out</u> and get it imprinted on your memory.

Population Growth and Structure

We're not done with the <u>DTM</u> I'm afraid, so keep that image of the graph clear in your mind.

A Country's Population Structure Changes with the Stages of the DTM

The <u>population structure</u> of a country is how many people there are of <u>each age group</u> in the population, and how many there are of <u>each sex</u>. It's shown using <u>population pyramids</u>.

The <u>changes</u> that happen between each stage of the <u>DTM</u> give countries <u>different</u> population structures:

Stage 1

Birth rate is <u>high</u> because there's <u>no</u> use of <u>contraception</u>, and people have <u>lots of children</u> because <u>many infants die</u>.

<u>Death rate</u> is <u>high</u> due to <u>poor healthcare</u>.

<u>Population growth rate</u> is <u>zero</u>.

<u>Population structure</u> — <u>life expectancy</u> is <u>low</u> (few people reach old age), so the population is made up of mostly <u>young people</u>.

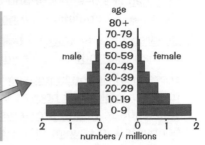

Stage 2

Birth rate is <u>high</u> because there's <u>no</u> use of <u>contraception</u>. Also, the economy is based on <u>agriculture</u> so people have <u>lots of children</u> to <u>work</u> on <u>farms</u>.

<u>Death rate falls</u> due to <u>improved healthcare</u>.

<u>Population growth rate</u> is <u>very high</u>.

<u>Population structure</u> — life expectancy has <u>increased</u>, but there are still <u>more young people</u> than <u>older people</u>.

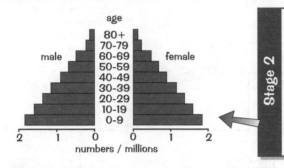

Stage 3

Birth rate is <u>rapidly falling</u> due to the <u>emancipation of women</u> (where they have a <u>more equal place</u> in society) and <u>better education</u>. The use of <u>contraception increases</u> and more women <u>work</u> instead of having children. The economy also changes to <u>manufacturing</u>, so fewer children are needed to work on farms.

<u>Death rate falls</u> due to more <u>medical advances</u>.

<u>Population growth rate</u> is <u>high</u>.

<u>Population structure</u> — <u>more people</u> are <u>living to be older</u>.

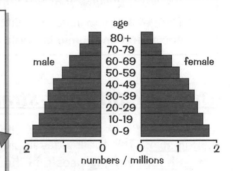

Stage 4

Birth rate is <u>low</u> — people move to <u>urban areas</u> (<u>urbanisation</u>), their <u>wealth improves</u> and they want <u>more possessions</u>. This means there's <u>less money</u> available for having children.

<u>Death rate</u> is <u>low and fluctuating</u>.

<u>Population growth rate</u> is <u>zero</u>.

<u>Population structure</u> — life expectancy is <u>high</u>, so even more people are living to be older.

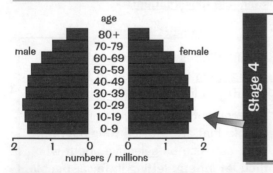

Stage 5

Birth rate is <u>slowly falling</u> — there's <u>less money</u> available to raise children because people have <u>dependent elderly relatives</u>.

<u>Death rate</u> is <u>low and fluctuating</u>.

<u>Population growth rate</u> is <u>negative</u>.

<u>Population structure</u> — <u>more older people</u> than young people.

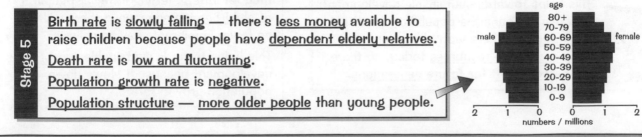

Population pyramids — no-one knows how the ancient people built them...

There are a fair few words on this page — but they're all important. You need to know all the <u>reasons</u> for the changes between the <u>DTM stages</u>, and the reasons for the changes in <u>population structure</u> too.

Managing Rapid Population Growth

Let's be clear here, <u>rapid population growth</u> doesn't mean everyone is getting taller and fatter. It's the rapid increase in the <u>number of people</u> in a country, and it can have some pretty <u>serious impacts</u>...

Rapid Population Growth has Social, Economic and Political Impacts

Social

1) <u>Services</u> like healthcare and education <u>can't cope</u> with the rapid increase in population, so <u>not everybody has access</u> to them.

2) <u>Children</u> have to <u>work</u> to help support their <u>large families</u>, so they <u>miss out</u> on their <u>education</u>.

3) There <u>aren't enough houses</u> for everyone, so people are forced to live in <u>makeshift houses</u> in <u>overcrowded settlements</u>. This leads to <u>health problems</u> because the houses aren't always connected to <u>sewers</u> or they don't have access to <u>clean water</u>.

4) There are <u>food shortages</u> if the country can't grow or import enough food for the population.

Economic

1) There <u>aren't enough jobs</u> for the number of people in the country, so <u>unemployment increases</u>.

2) There's <u>increased poverty</u> because more people are born into families that are <u>already poor</u>.

Now where's me pension gone?

Political

1) <u>Most</u> of the population is made up of <u>young people</u> so the government focuses on <u>policies</u> that are important to young people, e.g. <u>education</u> and provision of things like <u>childcare</u>.

2) There are <u>fewer older people</u> so the government <u>doesn't</u> have to focus on policies that are important to older people, e.g. <u>pensions</u>.

3) The government has to make policies to bring population growth <u>under control</u>, so the social and economic impacts of rapid population growth don't get <u>any worse</u>.

There are Different Strategies to Control Rapid Population Growth

Countries need to <u>control</u> rapid population growth and they also need to <u>develop</u> in a way that's <u>sustainable</u>. This means developing in a way that allows people <u>today</u> to get the things they need, but <u>without stopping</u> people in the <u>future</u> from getting what they <u>need</u>. Here are a couple of examples of <u>population policies</u> and how they help to achieve <u>sustainable development</u>:

Birth control programmes

Birth control programmes aim to <u>reduce</u> the <u>birth rate</u>.

Some governments do this by having <u>laws</u> about <u>how many children</u> couples are allowed to have (see the next page). Governments also help couples to <u>plan</u> (and <u>limit</u>) how many children they have by offering <u>free contraception</u> and <u>sex education</u>.

This helps towards sustainable development because it means the population won't get <u>much bigger</u>. There won't be many more people <u>using up resources</u> today, so there will be <u>some left</u> for <u>future generations</u>.

Immigration laws

Immigration laws aim to <u>control</u> <u>immigration</u> (people moving to a country to live there <u>permanently</u>).

Governments can <u>limit</u> the <u>number</u> of people that are allowed to immigrate. They can also be <u>selective</u> about who they let in, e.g. letting in <u>fewer</u> people of <u>child-bearing age</u> means there will be fewer immigrants having children.

This helps towards sustainable development because it <u>slows down</u> population growth rate.

When it comes to population growth I blame the parents...

It's not always a case of 'the more the merrier' — rapid population growth can cause <u>problems</u>. That's where <u>strategies</u> to <u>control</u> population growth come in. You need to know them and how they relate to <u>sustainable development</u>.

Managing Population Growth — Case Studies

This topic would be sad and lonely without <u>case studies</u>...

China has a Strict Birth Control Programme

1) China has the <u>largest</u> population of any country in the world — over <u>1.3 billion</u>.

2) Different <u>policies</u> have been used to <u>control</u> rapid population growth — the most important is the '<u>one-child policy</u>' introduced in <u>1979</u>. This means that all couples are <u>very strongly encouraged</u> to have <u>only one child</u>.

3) Couples that only have one child are given <u>benefits</u> like <u>longer maternity leave</u>, <u>better housing</u> and <u>free education</u> for the child. Couples that have more than one child <u>don't get</u> any benefits and are also <u>fined</u> part of their income.

4) Over the years, the policy has <u>changed</u> so there are some exceptions:

- In some <u>rural areas</u>, couples are <u>allowed</u> to have a <u>second child</u> if the first is a <u>girl</u>, or has a <u>physical disability</u>. This is because more children are still <u>needed</u> to <u>work</u> on <u>farms</u> in rural areas.

- If one of the parents has a <u>disability</u> or if both parents are <u>only children</u>, then couples are allowed to have a second child. This is so there are enough people to <u>look after</u> the parents.

Effectiveness	1) The policy has <u>prevented</u> up to <u>400 million births</u>. The <u>fertility rate</u> (the average number of children a woman will have in her life) has <u>dropped</u> from <u>5.7</u> in 1970 to around <u>1.8</u> today.	2) Some people think that it <u>wasn't just</u> the one-child policy that slowed population growth. They say <u>older policies</u> about leaving <u>longer gaps between children</u> were <u>more effective</u>, and that Chinese people <u>want fewer children</u> anyway as they've become <u>more wealthy</u>.

China's one-child policy <u>helps</u> towards <u>sustainable development</u> — the population hasn't <u>grown as fast</u> (and got as <u>big</u>) as it would have done <u>without</u> the policy, so <u>fewer resources</u> have been <u>used</u>.

Indonesia has Tried to Tackle the Problems of Rapid Population Growth

1) Indonesia is a country made up of <u>thousands of islands</u>. It has the <u>fourth largest</u> population of any country in the world — over <u>240 million</u>.

2) The population <u>isn't distributed evenly</u> — most people (around <u>130 million</u>) live on the island of <u>Java</u>.

3) This has led to <u>social</u> and <u>economic problems</u> (see the previous page) on the <u>densely</u> populated islands, e.g. a <u>lack</u> of adequate <u>services</u> and <u>housing</u> as well as <u>unemployment</u> and <u>poverty</u>.

4) The Indonesian Government started a policy in the 1960s called the <u>transmigration policy</u>, which aims to reduce the <u>impacts</u> of population growth.

5) <u>Millions</u> of people have been <u>moved</u> from the <u>densely</u> populated islands like Java, to the <u>less densely</u> populated islands like <u>Sumatra</u>.

☐ = Indonesia

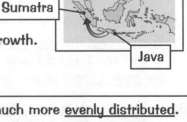

Sumatra

Java

Effectiveness	1) <u>Millions</u> of people have been moved, but the population still <u>isn't</u> much more <u>evenly distributed</u>.
	2) Not all the people who were moved <u>escaped poverty</u> — either they <u>didn't</u> have the <u>skills</u> to farm the land, or the <u>land</u> was <u>too poor</u> to be farmed on their new island.
	3) Lots of people were moved to land that was <u>already occupied</u> by <u>native people</u>. This created a <u>new problem</u> — <u>conflict</u> between the natives and the migrants.

Indonesia's transmigration policy <u>hasn't helped</u> towards <u>sustainable development</u> because it only reduces the <u>impacts</u> of population growth — the population is still getting <u>much bigger</u>.

Someone should tell the pandas they can have as many babies as they like...

In some cases, governments have had to take pretty <u>drastic action</u> to deal with <u>rapid population growth</u>. Just try and imagine what you'd say if your local MP knocked on your door and told you to <u>move</u> to another island...

Managing Ageing Populations

An <u>ageing population</u> is one that has a <u>high proportion</u> of <u>older people</u>. Ageing populations can face some sticky <u>economic</u> and <u>social</u> problems. They also tend to consume a high quantity of teacakes.

An Ageing Population Impacts on Future Development

The <u>population structure</u> of an ageing population has <u>more</u> <u>older</u> people than <u>younger</u> people because few people are being <u>born</u>, and more people are <u>surviving</u> to old age.

Countries with an ageing population are usually the <u>richer countries</u> in Stage <u>5</u> of the <u>DTM</u> (see p. 76).

Older people (over 65) are <u>supported</u> by the <u>working population</u> (aged 16-64) — they're <u>dependent</u> on them. So in a country with an ageing population there's a <u>higher proportion</u> of people who are dependent. This has <u>economic</u> and <u>social impacts</u>, which can affect a country's <u>future development</u>:

age: 80+, 70-79, 60-69, 50-59, 40-49, 30-39, 20-29, 10-19, 0-9
male | female
2 1 0 0 1 2
numbers / millions

ECONOMIC

1) The working population pay <u>taxes</u>, some of which the government use to pay the <u>state pensions</u> of older people, and to pay for <u>services</u> like retirement homes and healthcare. Taxes would need to <u>go up</u> because there are <u>more pensions</u> to pay for, and older people need <u>more healthcare</u>.

2) The <u>economy</u> of the country would <u>grow more slowly</u> — <u>less money</u> is being spent on things that help the economy to <u>grow</u>, e.g. education and business, and <u>more money</u> is being spent on things that <u>don't</u> help the economy to grow, e.g. retirement homes.

SOCIAL

1) <u>Healthcare services</u> are <u>stretched more</u> because older people need more medical care.

2) People will need to spend more time working as <u>unpaid carers</u> for older family members. This means that the working population have <u>less</u> <u>leisure time</u> and are more <u>stressed</u> and <u>worried</u>.

3) People may have <u>fewer children</u> because they <u>can't afford</u> lots of children when they have dependent older relatives. This leads to a <u>drop in birth rate</u>.

4) The <u>more</u> old people there are, the <u>lower</u> the <u>pension</u> provided by the government will be. People will have to <u>retire later</u> because they <u>can't afford</u> to get by on a state pension.

There are Different Strategies to Cope with an Ageing Population

1) <u>Encouraging larger families</u>, e.g. in Italy women are offered <u>cash rewards</u> to have more children. This <u>increases</u> the <u>number of young people</u> — when they start work there will be a <u>larger working population</u> to pay taxes and support the ageing population.

2) <u>Encouraging</u> the <u>immigration</u> of <u>young people</u> from other countries. This <u>increases</u> the <u>working population</u> so there are <u>more people</u> paying taxes to support the ageing population.

> These strategies <u>don't help</u> towards <u>sustainable development</u> because they <u>increase</u> the <u>population size</u>.

3) <u>Raising</u> the <u>retirement age</u> — people <u>stay in work longer</u> and contribute to state pensions and personal pensions for <u>longer</u>. They will also <u>claim</u> the <u>state pension</u> for <u>less time</u>.

4) <u>Raising taxes</u> for the working population — this would <u>increase</u> the amount of <u>money available</u> to support the ageing population.

> These strategies <u>help</u> towards <u>sustainable development</u> because they help to <u>reduce the impacts</u> of an ageing population, <u>without increasing</u> the <u>population size</u>.

What are taxes, Grandpa?

You'll see, Jimmy. You'll see...

Being old is not a crime but my great-aunt has knitted some criminal jumpers...

'Live long enough to be a burden on your children' — I thought it was just a phrase... Learn about the <u>social</u> and <u>economic impacts</u> an ageing population can have, and the <u>strategies</u> governments have come up with to deal with them.

Ageing Populations — Case Study

Like most wealthy and developed countries around the world, the UK has an ageing population. This case study's got lots of <u>juicy statistics</u> you can use in your exam answers, so get <u>swotting</u>.

The UK's Population is Ageing

In 2005, <u>16%</u> of the population of the UK was <u>over 65</u>. By 2041 this could be <u>25%</u>.

The Ageing Population is Caused by Increasing Life Expectancy and Dropping Birth Rate

1) <u>People are living longer</u> because of advances in <u>medicine</u> and improved <u>living standards</u>. Between 1980 and 2006 <u>life expectancy rose</u> 2.6 years for women and 6.4 years for men — it's currently <u>81.5</u> for women and <u>77.2</u> for men. This means the <u>proportion</u> of older people in the population is <u>going up</u>.

2) <u>Lots of babies</u> were born in the <u>1940s</u> and <u>1960s</u> — periods called '<u>baby booms</u>'. Those born in the 1940s are <u>retiring</u> now, creating a '<u>pensioner boom</u>'.

3) Since the 1970s, the number of <u>babies born</u> has <u>fallen</u>. With <u>fewer young people</u> in the population the proportion of older people <u>goes up</u>.

UK population pyramid — age: 80+, 70-79, 60-69, 50-59, 40-49, 30-39, 20-29, 10-19, 0-9; male / female; numbers / millions 5 4 3 2 1 0 0 1 2 3 4 5

The UK's Ageing Population Causes a few Problems

1) <u>More elderly people</u> are living in <u>poverty</u> — the working population <u>isn't large enough</u> to pay for a decent pension, and many people <u>don't have other savings</u>.

2) Even though the <u>state pension</u> is low the government is <u>struggling</u> to <u>pay it</u>. The taxes paid by people in work <u>aren't enough</u> to cover the cost of pensions and as the population ages the situation is <u>getting worse</u>.

3) The <u>health service</u> is <u>under pressure</u> because older people need more medical care than younger people. For example, in 2005 the <u>average stay</u> in hospital for people over 75 was <u>13 nights</u>, but for the whole of the UK the average stay was only <u>8 nights</u>.

The UK Government has Strategies to Cope with the Ageing Population

1) <u>Raise the retirement age</u> — the retirement age in the UK is currently <u>65</u> for men and <u>60</u> for women. This is going to change in stages, so that by <u>2046</u> it will be <u>68</u> for everyone. People will have to <u>work for longer</u>, so there will be <u>more people paying tax</u> and <u>fewer claiming</u> a pension.

2) <u>Encourage immigration of young people to the UK</u> — the UK has <u>allowed immigration</u> of people from countries that joined the EU in 2004. Around <u>80%</u> of immigrants from new EU countries in 2004 were <u>34 or under</u>. This increases the number of people <u>paying taxes</u>, which helps to pay for the <u>state pension</u> and <u>services</u>.

3) <u>Encourage women to have children</u> — working family <u>tax credits</u> support women (and men) who go back to <u>work</u> after their children are born. This makes it <u>more affordable</u> for couples to <u>have children</u>.

4) <u>Encourage people to take out private pensions</u> — the government gives <u>tax breaks</u> for some types of private pension. With private pensions, people <u>won't</u> be so <u>dependent</u> on the state pension.

The Blurb STATE PENSION AGE RAISED TO 105

We Don't Know if the Strategies have Worked Yet

It's <u>too early</u> to tell if government strategies are working. Even if they do have some effect it's likely that <u>future generations</u> will have to <u>work longer</u> and <u>rely on their families</u> to support them in old age.

Ageing population impact — sales of toilet roll covers are through the roof...

This case study is perfect exam fodder. Nothing makes an examiner's eyes spin like fruit machines more than <u>real-life examples</u>. Memorise the <u>facts</u> and <u>figures</u> from this page and you'll be all set up to get top marks in your exam.

Unit 2A — Population Change

Population Movements

Migration is the movement of people from one area to another area. Alas, they don't just do it for the fun of it — you need to learn the reasons why, as well as some good ol' impacts.

People Migrate Within Countries and To Different Countries

1) When people move into an area, it's called immigration. The people are called immigrants.

2) When people leave an area, it's called emigration. The people are called emigrants.

3) People can move to different countries. This can be across the world, or just a few miles over a border.

4) People can move between different regions within countries, e.g. from the countryside to a city (called rural-urban migration).

Migration Happens Because of Push and Pull Factors

The reasons a person migrates can be classified as either push or pull factors:

Push factors are the things about a person's place of origin (where they originally lived) that make them decide to move.

Pull factors are things about a person's destination that attracts them.

EXAMPLE PUSH FACTORS:

They're usually negative things like not being able to find a job, poor living conditions, war or a natural disaster in their country of origin.

For example, refugees are people who've been forced to leave their country due to war or a natural disaster, e.g. thousands of refugees migrated to escape the war in Kosovo in 1999.

EXAMPLE PULL FACTORS:

They're usually positive things such as job opportunities or a better standard of living.

For example, economic migrants are people who move voluntarily from poor places to richer places looking for jobs or higher wages, e.g. from Mexico to the USA. They often migrate so they can earn more money and then send some back to family in their country of origin.

Migration Has Positive and Negative Impacts

Migration has impacts on both the source country (where they come from) and the receiving country (where they're going to):

	Positive impacts	Negative impacts
Source country	Reduced demand on services, e.g. schools and hospitals. Money is sent back to the source country by emigrants.	Labour shortage — it's mostly people of working age that emigrate. Skills shortage — sometimes it's the more highly educated people that emigrate. Ageing population — there's a high proportion of older people left.
Receiving country	Increased labour force — young people immigrate to find work. Migrant workers pay taxes that help to fund services.	Locals and immigrants compete for jobs — this can lead to tension and even conflict. Increased demand for services, e.g. overcrowding in schools and hospitals. Not all the money earnt by immigrants is spent in the destination country — some is sent back to their country of origin.

Umm-igration is when you don't know whether you're coming or going...

Migration sounds like a rough business, people being pushed and pulled all over the place. You need to remember that migration affects both the place the people leave, and the place they go to, and the effects can be positive or negative.

Migration Within and To the EU

People move around <u>within</u> the EU <u>voluntarily</u> looking for <u>better paid work</u>. Refugees move <u>to the EU</u> because they've been <u>forced out</u> of their country. Here are a couple of <u>examples</u> to spice up your life...

There are Economic Migrations Within the EU

People who come from a country in the EU can <u>live</u> and <u>work</u> in <u>any other</u> EU country. In 2004, ten eastern European countries joined the EU. Since then, people from these countries have been moving to other EU countries. More than <u>half a million</u> people from Poland came <u>to the UK</u> between <u>2004</u> and <u>2007</u>.

There were <u>push and pull factors</u> for why people left Poland and came to the UK:

<u>Push factors</u> from Poland (in 2004):

1) <u>High unemployment</u> — around <u>19%</u>.

2) <u>Low average wages</u> — about <u>one third</u> of the average EU wage.

3) <u>Housing shortages</u> — just over <u>300</u> dwellings for every <u>1000 people</u>.

<u>Pull factors</u> to the UK:

1) <u>Ease of migration</u> — the UK allowed <u>unlimited migration</u> in 2004 (it was <u>restricted</u> in some other EU countries).

2) <u>More work and higher wages</u> — wages in the UK were <u>higher</u> and there was a <u>big demand</u> for <u>tradesmen</u>, e.g. plumbers.

3) <u>Good exchange rate</u> — the <u>pound</u> was <u>worth a lot</u> of <u>Polish currency</u>, so sending a few pounds back to Poland made a <u>big difference</u> to family at home.

IMPACTS IN POLAND

1) Poland's <u>population fell</u> (by 0.3% between 2003 and 2007), and the <u>birth rate fell</u> as most people who left were <u>young</u>.

2) There was a <u>shortage of workers</u> in Poland, <u>slowing</u> the <u>growth</u> of the <u>economy</u>.

3) The Polish <u>economy</u> was <u>boosted</u> by the money <u>sent home</u> from emigrants — around <u>€3 billion</u> was sent to Poland from abroad in 2006.

IMPACTS IN THE UK

1) The UK <u>population went up slightly</u>.

2) Immigration <u>boosted</u> the UK <u>economy</u>, but a lot of the money earned in the UK was <u>sent home</u>.

3) <u>New shops</u> selling Polish products opened to serve new Polish communities.

4) Many Poles are <u>Catholic</u> so <u>attendance</u> at Catholic <u>churches went up</u>.

Refugees Migrate To the EU

Huge numbers of people migrate from <u>Africa to the EU</u>. For example, by crossing the Mediterranean sea to <u>Spain</u> — in 2001, 45 000 emigrants from Africa were caught and refused entry to Spain.

Many of these migrants are <u>refugees</u> (see the previous page) from <u>wars</u> in central and western African countries. For example, more than <u>2 million</u> people were <u>forced from their homes</u> because of the civil war in Sierra Leone (in West Africa) between 1991 and 2002.

There are only <u>push factors</u> for African refugees of war — people flee the countries because of the threat of <u>violence</u> or <u>death</u> during the wars.

Here are some of the <u>impacts</u>:

IMPACTS IN AFRICAN COUNTRIES

1) The <u>working population</u> is <u>reduced</u> so there are <u>fewer people</u> <u>contributing</u> to the <u>economy</u>.

2) <u>Families become separated</u> when fleeing from wars.

IMPACTS IN SPAIN

1) <u>Social tension</u> between immigrants and Spaniards.

2) <u>More unskilled workers</u> in Spain, which has filled <u>gaps</u> in the labour market.

3) <u>Average wages</u> for unskilled jobs have <u>fallen</u> because there are <u>so many</u> people who want the jobs.

4) The <u>birth rate</u> has <u>increased</u> because there are so many <u>young</u> immigrants.

Economic migration — the money's getting away...

There's plenty of migration going on <u>within</u> and <u>into</u> the EU. Never mind whether British beaches are rubbish, French toilets are awful and German pop music is inexcusable — looks like the EU's the place to be right now.

Revision Summary for Population Change

That's another smashing topic under your belt — congratulations, oh studious one. And here's a delightful array of questions so you can check you've taken it all in. If you'd care to begin...

1) What are the two things that affect the population size of the world?

2) Under what circumstances does natural increase happen to a population?

3) 'The world population growth rate is increasing exponentially'. What does this mean?

4) What happens to death rate at Stage 2 of the DTM?

5) What happens to birth rate at Stage 3 of the DTM?

6) Are richer countries or poorer countries more likely to be in the early stages of the DTM?

7) Give one reason why birth rate is high during Stage 1 of the DTM.

8) Describe how changes in the economy affect the population growth rate.

9) Give one reason why the birth rate rapidly falls during Stage 3 of the DTM.

10) Briefly describe the population structure of a country in Stage 5 of the DTM.

11) Give two social impacts of rapid population growth.

12) Give one economic impact of rapid population growth.

13) Describe what it means for a country to develop in a way that's sustainable.

14) Give an example of a strategy a country could use to control rapid population growth.

15) a) Name a birth control population policy.

    b) Describe the policy.

    c) Give one piece of evidence that shows the policy has been effective.

16) a) Name a non-birth control population policy.

    b) Describe the policy.

    c) Give one reason why the policy hasn't been effective.

17) What is an ageing population?

18) Give two causes of an ageing population.

19) Give one economic impact of an ageing population.

20) Give two social impacts of an ageing population.

21) Describe one strategy to cope with an ageing population.

22) Describe one problem caused by the UK's ageing population.

23) Give one strategy used by the UK Government to cope with an ageing population.

24) Define 'migration'.

25) What's it called when a person moves into an area?

26) a) Explain what 'pull factors' are.

    b) Give an example of a pull factor.

27) Give one negative impact of migration on a source country.

28) Give one positive impact of migration on a receiving country.

29) a) Describe an example of economic migration within the EU.

    b) Give one impact on the source country, and one impact on the receiving country.

30) a) Describe an example of refugee migration to the EU.

    b) Describe the push factors.

Urbanisation

Urban areas (towns and cities) are the place to be... or so I'm told by my mate who tucks his tracksuit bottoms into his socks. People are upping sticks and moving to cities, and you've got to know all about it.

Urbanisation is Happening Fastest in Poorer Countries

Urbanisation is the growth in the proportion of a country's population living in urban areas. It's happening in countries all over the world — more than 50% of the world's population currently live in urban areas (3.4 billion people) and this is increasing every day. But urbanisation differs between richer and poorer countries:

1) Most of the population in richer countries already live in urban areas, e.g. more than 80% of the UK's population live in urban areas.

2) Not many of the population in poorer countries currently live in urban areas, e.g. around 25% of the population of Bangladesh live in urban areas.

3) Most urbanisation that's happening in the world today is going on in poorer countries and it's happening at a fast pace.

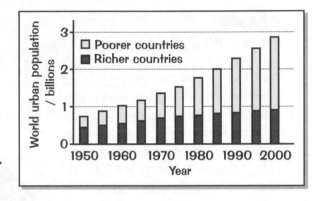

Urbanisation is Caused By Rural-urban Migration...

Rural-urban migration is the movement of people from the countryside to the cities. Rural-urban migration causes urbanisation in richer and poorer countries. The reasons why people move are different in poorer and richer countries though.

Here are a couple of reasons why people in poorer countries move from rural areas to cities:

1) There's often a shortage of services (e.g. education, access to water and power) in rural areas. Also, people from rural areas sometimes believe that the standard of living is better in cities (even though this often turns out not to be the case).

2) There are more jobs in urban areas. Industry is attracted to cities because there's a larger workforce and better infrastructure than in rural areas.

Here are a couple of reasons why people in richer countries move from rural areas to cities:

1) Most urbanisation in rich countries occurred during the Industrial and the Agricultural Revolutions (18th and 19th centuries) — machinery began to replace farm labour in rural areas, and jobs were created in new factories in urban areas. People moved from farms to towns for work.

2) In the late 20th century, people left run-down inner city areas and moved to the country. But people are now being encouraged back by the redevelopment of these areas.

... And Good Healthcare and a High Birth Rate in Cities

It's normally young people that move to cities to find work. These people have children in the cities, which increases the proportion of the population living in urban areas. Also, better healthcare in urban areas means people live longer, again increasing the proportion of people in urban areas.

Dick Whittington and His Cat — the classic rural-urban migration case study...

Nothing too difficult on this page — richer countries have a high percentage of their population in urban areas, but urbanisation in poorer countries is happening faster than you can say 'oh my giddy aunt, that's some rapid urbanisation'.

Parts of a City

Every city is different, but they all have <u>dodgy run-down parts</u> and <u>posh housing areas</u>. Obviously you should use more <u>formal terms</u> to describe them in the exam — amazingly enough, they're all listed below...

A City can be Split into Four Main Parts

Cities are usually made up of <u>four parts</u> — each part has a <u>different land use</u> (e.g. housing or industrial). The diagram below is a <u>view from above</u> of a <u>typical city</u> — it shows <u>roughly</u> where the four parts are. The land use of each part <u>stays fairly similar</u> from <u>city to city</u> — the diagram below shows what the land use is in a city in a <u>rich country</u>, but it <u>can differ</u> a bit (see below):

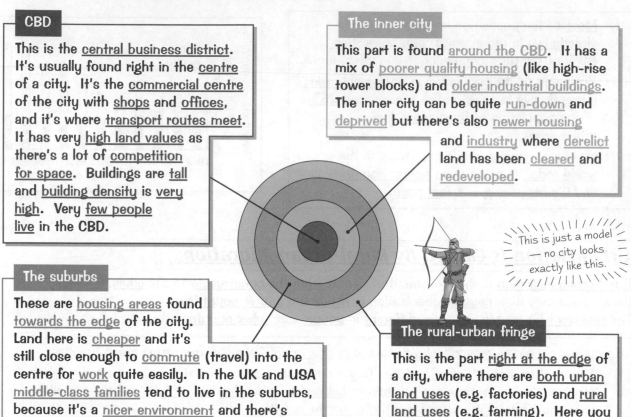

CBD

This is the <u>central business district</u>. It's usually found right in the <u>centre</u> of a city. It's the <u>commercial centre</u> of the city with <u>shops</u> and <u>offices</u>, and it's where <u>transport routes meet</u>. It has very <u>high land values</u> as there's a lot of <u>competition for space</u>. Buildings are <u>tall</u> and <u>building density</u> is <u>very high</u>. Very <u>few people live</u> in the CBD.

The inner city

This part is found <u>around the CBD</u>. It has a mix of <u>poorer quality housing</u> (like high-rise tower blocks) and <u>older industrial buildings</u>. The inner city can be quite <u>run-down</u> and <u>deprived</u> but there's also <u>newer housing</u> and <u>industry</u> where <u>derelict</u> land has been <u>cleared</u> and <u>redeveloped</u>.

The suburbs

These are <u>housing areas</u> found <u>towards the edge</u> of the city. Land here is <u>cheaper</u> and it's still close enough to <u>commute</u> (travel) into the centre for <u>work</u> quite easily. In the UK and USA <u>middle-class families</u> tend to live in the suburbs, because it's a <u>nicer environment</u> and there's <u>less crime</u> and <u>pollution</u> than the inner city.

The rural-urban fringe

This is the part <u>right at the edge</u> of a city, where there are <u>both urban land uses</u> (e.g. factories) and <u>rural land uses</u> (e.g. farming). Here you tend to find <u>fewer</u>, <u>larger houses</u>.

This is just a model — no city looks exactly like this.

The Land Use of the Parts can Differ from City to City

1) Sometimes the <u>land use</u> of each part <u>doesn't match the model</u> above — real cities are all <u>slightly different</u>. For example, in countries like <u>France</u>, <u>Italy</u> and <u>Sweden</u>, the <u>inner city</u> areas are where the <u>wealthier middle-classes</u> live and the <u>suburbs</u> tend to be the more <u>deprived</u> areas.

2) The <u>land use</u> of each part of a city can also <u>change over time</u>, for example:

   - In recent years a lot of <u>shopping centres</u> have been built in <u>out-of-town</u> locations in the UK (which has caused <u>shops in CBDs</u> to <u>close down</u>).
   - <u>Inner city tower blocks</u> have been <u>removed</u> and <u>replaced</u> with <u>housing estates</u> on the <u>rural–urban fringe</u>.
   - <u>New housing</u> is often built on <u>brownfield sites</u> (cleared derelict land) in the <u>inner city instead of</u> toward the <u>edges</u> of the city.

Just popping into the rural-urban fringe to do a bit of shopping...

You need to know about the <u>four main parts</u> of a city and the <u>land use</u> in each bit, but remember that the land use <u>isn't the same everywhere</u> — I'm sure city planners do this on purpose, just to make your revision awkward. Some people...

Urban Issues

Not everything is tickety-boo in underline urban areas in underline richer countries — they have underline social and underline environmental underline problems. Since you asked nicely, you can learn all about them...

Many Urban Areas Have the Same Problems

Cities in underline richer countries all have the underline same kind of problems:

1 A underline shortage of good quality underline housing.   **2** underline Run down CBDs.   **3** underline Traffic congestion and underline pollution from underline cars.

4 underline Ethnic segregation (people from different races and religions not mixing)

Over the underline next couple of pages you'll look at underline each problem and the underline solutions for them in a underline bit more detail.

Growing Populations Need More Housing

Some richer countries (e.g. the UK) have underline housing shortages in underline urban areas because the underline urban population has underline grown quickly. Here are a few ways the shortages are being underline tackled:

1) underline Urban renewal schemes

- These are underline government strategies first widely used in the 1990s. They underline encourage underline investment in underline new housing, underline services and underline employment in underline derelict inner city areas.
- A successful example is the underline dockland development in underline Liverpool — the derelict docks (a brownfield site) were converted into underline high quality housing with underline good local services.

2) underline New towns

- underline Brand new towns have been built to house the underline overspill populations from existing towns and cities where there was a underline shortage of housing. Milton Keynes the one of the most well-known examples of a underline new town — building started in 1970.

3) underline Relocation incentives

- These are used to underline encourage people living in underline large council houses (who underline don't really need a underline big house or to underline live in the city) to underline move out of urban areas. This underline frees up underline houses in urban areas for other people, e.g. underline working families.
- For example, a scheme that's run by a underline London council encourages underline older people who live in underline big houses in the city to underline move to the underline seaside or the underline countryside. The council underline helps people who volunteer to underline move out and it also underline gives them money.

Efforts are being Made to Revitalise CBDs

The underline CBDs in some cities are underline run down. One reason for this is competition from underline out-of-town shopping centres and underline business parks, which have underline cheaper rent (so underline lure shops to move there) and underline are easier to underline drive to. But steps are being taken to underline revitalise some CBDs and underline attract people back to them. For example:

1) underline Pedestrianising areas (stopping car access) to make them underline safer and underline nicer for shoppers.
2) underline Improving access with underline better public transport links and underline better car parking.
3) Converting derelict warehouses and docks into smart new shops, restaurants and museums.
4) Improving public areas, e.g. parks and squares, to make them more attractive.

Initial government investment encourages businesses to return, attracting more customers, which attracts more businesses and so on. The London docklands development is a good example of this.

Sittin' on the dock of the bay — in a very chic new Italian place...

A lack of housing and run down CBDs — it's easy to see why people love cities. To be fair, those city slickers have some good ideas on how to deal with housing shortages and revitalise CBDs — make sure you know all about them.

Urban Issues

If you're hungry for more <u>urban problems</u> and <u>solutions</u> then you've come to the right place. Bon appétit...

Increased Car Use has an Impact on Urban Environments

There are <u>more and more cars</u> on the roads of cities in richer countries. This causes
a variety of <u>problems</u>, which can <u>discourage</u> people from visiting and shopping in the city:

1 More air pollution,
which damages health

2 More road
accidents

3 Air pollution also
damages buildings

4 More traffic jams
and congestion

There are a variety of <u>solutions</u> to help <u>reduce traffic</u> and its <u>impacts</u>:

1) <u>Improving public transport</u>. This <u>encourages</u> people to <u>use public transport instead of cars</u>,
which <u>reduces traffic congestion</u>, <u>air pollution</u>, <u>traffic jams</u> and <u>accidents</u>.

2) <u>Increasing car parking charges</u> in city centres. This <u>discourages car use</u>, so people are
<u>more likely to use public transport</u> instead.

3) <u>Bus priority lanes</u> — these <u>speed up bus services</u> so people are <u>more likely to use them</u>.

4) <u>Pedestrianisation</u> of central areas. This <u>removes traffic</u> from the main shopping streets, which <u>reduces</u>
the number of <u>accidents</u> and <u>pollution levels</u>. It also makes these areas <u>more attractive to shoppers</u>.

Many Urban Areas Have a Variety of Cultures

Cities usually have a <u>variety</u> of people from different <u>ethnic backgrounds</u> (people from different <u>races</u>
and <u>religions</u>). But there's often <u>ethnic segregation</u> in urban areas, i.e. people of different ethnicities
not mixing. There are several <u>reasons</u> for this:

1) People <u>prefer</u> to <u>live close</u> to others with the <u>same background</u> and <u>religion</u>,
and who speak the <u>same language</u>.

2) People <u>live near</u> to <u>services</u> that are <u>important</u> to <u>their culture</u>, e.g. <u>places of worship</u>.
This means people of the <u>same ethnic background</u> tend to <u>live</u> in the <u>same area</u>.

3) People from the <u>same ethnic background</u> are often <u>restricted</u> in where they can live in
the <u>same way</u>, e.g. because of a <u>lack of money</u>, so they all end up in the <u>same place</u>.

Strategies to <u>support</u> the multicultural nature of urban areas <u>aren't</u> aimed at <u>forcing</u> people to <u>mix</u>.
The strategies make sure that everyone has <u>equal access to services</u>, like <u>healthcare</u> and <u>education</u>.
Some of the ways to do this include:

1) Making sure <u>everyone</u> can <u>access information</u> about the different <u>services</u>,
e.g. by printing leaflets in a variety of languages.

2) <u>Improving communication</u> between all parts of the community, e.g. by <u>involving</u>
the <u>leaders</u> of <u>different ethnic communities</u> when <u>making decisions</u>.

3) <u>Providing interpreters</u> at places like hospitals and police stations.

4) Making sure there are <u>suitable services</u> for the <u>different cultures</u>. For example, in <u>some cultures</u>
it's <u>unacceptable</u> to be <u>seen by a doctor</u> of the <u>opposite sex</u>, so <u>alternatives</u> should be <u>provided</u>.

Perhaps it would also help if the slang term for bus wasn't 'loser cruiser'...

All this stuff seems pretty straightforward but keep going over it until it's all lodged in your noggin. You should be
able to <u>describe</u> the <u>problems and the solutions</u> — you could be asked a question about either (or both of them).

Squatter Settlements

If you think your living room doesn't have enough comfy sofas, or your bathroom's just not big enough, then spare a thought for people who live in <u>squatter settlements</u> — they're not exactly plush places to live.

Squatter Settlements are Common in Cities in Poorer Countries

1) <u>Squatter settlements</u> are settlements that are built <u>illegally</u> in and around the city, by people who <u>can't afford proper housing</u>.

2) Squatter settlements are a problem in many <u>growing</u> cities in <u>poorer countries</u>, e.g. <u>São Paulo</u> (Brazil) and <u>Mumbai</u> (India).

3) Most of the inhabitants have <u>moved to the city</u> from the <u>countryside</u> — they're <u>rural–urban migrants</u>.

4) The settlements are <u>badly built</u> and <u>overcrowded</u>. They often <u>don't</u> have <u>basic services</u> like <u>electricity</u> or <u>sewers</u>.

5) They're called <u>favelas</u> in Brazil and <u>shanty towns</u> or <u>slums</u> in some other places.

Little space between houses

No electricity or phone lines

Houses built from waste material like plastic sheets

©iStockphoto.com/Nitin Sanil

No paved roads or sewers

Life in a squatter settlement can be <u>hard</u> and <u>dangerous</u> — the people living there <u>don't</u> have access to basic <u>services</u> like clean <u>running water</u>, proper <u>sewers</u> or <u>electricity</u>. They may also lack <u>policing</u>, <u>medical services</u> and <u>fire fighting</u>. Because of these problems, <u>life expectancy</u> is often <u>lower</u> than in the main city. Many inhabitants <u>work within the settlements</u>, e.g. in factories and shops. The jobs <u>aren't taxed</u> or <u>monitored</u> by the <u>government</u> — they're referred to as the <u>informal sector</u> of the economy. People often work <u>long hours</u> for <u>little pay</u> in the informal sector. But squatter settlements often govern themselves <u>more successfully</u> than you might expect and have a <u>strong community spirit</u>.

There are Ways to Improve Squatter Settlements

Squatter settlements <u>aren't great places to live</u>, but there's often <u>nowhere else for poor migrants to go</u>. People living in squatter settlements usually try to <u>improve</u> the settlements themselves. For example, neighbours <u>help each other</u> with <u>building</u> and some have even built <u>small schools</u>. But the residents have <u>little money</u> and can achieve <u>much more</u> with a bit of <u>help</u>:

SELF-HELP SCHEMES

These involve the <u>government</u> and <u>local people</u> <u>working together</u> to <u>improve life</u> in the settlement. The government <u>supplies building materials</u> and local people use them to <u>build their own homes</u>. This helps to provide <u>better housing</u> and the money saved on labour can be used to <u>provide</u> <u>basic services</u> like electricity and sewers.

SITE AND SERVICE SCHEMES

People <u>pay</u> a small amount of <u>rent</u> for a site, and they can <u>borrow money</u> to buy building materials to <u>build</u> or <u>improve</u> a <u>house</u> on their plot. The rent money is then used to <u>provide</u> <u>basic services</u> for the area. An example is the <u>Dandora scheme</u> in Nairobi, Kenya.

LOCAL AUTHORITY SCHEMES

These are <u>funded by the local government</u> and are about <u>improving</u> the <u>temporary accommodation</u> built by residents. For example, the City of Rio (Brazil) spent <u>$120 million</u> on the <u>Favela-Bairro project</u>, which aimed to <u>improve life</u> for the inhabitants of Rio de Janeiro's favelas (see next page).

Some residents even try to win 'Who Wants to be a Millionaire'...

People who live in <u>squatter settlements</u> try to improve the conditions <u>themselves</u>. It's not much like your dad's failed DIY attempts though — forget about putting up wonky shelves, these people are putting a roof over their heads.

Squatter Settlements — Case Study

And now, introducing tonight's <u>case study</u>, all the way from Brazil, lets hear a big hand for the <u>Favela-Bairro Project</u> in Rio de Janeiro. Oh come on, surely you can muster a smile...

The Favela-Bairro Project Helps People in Rio de Janeiro's Favelas

1) <u>Rio de Janeiro</u> is in south east <u>Brazil</u>. It has <u>600</u> squatter settlements (favelas), housing <u>one-fifth</u> of the city's population (more than <u>one million people</u>).

2) The <u>Favela-Bairro project</u> started in <u>1995</u> and is so <u>successful</u> it's been suggested as a <u>model</u> for <u>redeveloping other squatter settlements</u>.

3) The project involves <u>253 000 people</u> in <u>73 favelas</u>, and is being <u>extended</u> to help <u>even more people</u>.

4) 40% of the <u>$300 million funding</u> for the project came from the <u>local authority</u>. The rest was provided by an <u>international organisation</u> called the <u>Inter-American Development Bank</u>.

The Project Includes Social, Economic and Environmental Improvements

1) <u>Social</u> improvements:

- <u>Daycare centres</u> and <u>after school schemes</u> to <u>look after children</u> while their parents <u>work</u>.
- <u>Adult education classes</u> to <u>improve adult literacy</u>.
- Services to help people affected by <u>drug addiction</u>, <u>alcohol addiction</u> and <u>domestic violence</u>.

2) <u>Economic</u> improvements:

- Residents can now apply to <u>legally own</u> their properties.
- <u>Training schemes</u> to help people <u>learn new skills</u> so they can <u>find better jobs</u> and <u>earn more</u>.

3) <u>Environmental</u> improvements:

- <u>Replacement</u> of <u>wooden</u> buildings with <u>brick</u> buildings and the <u>removal</u> of homes on <u>dangerously steep slopes</u>.
- <u>Widening</u> and <u>paving</u> of streets to allow <u>easier access</u> (especially for <u>emergency services</u>).
- Provision of <u>basic services</u> such as <u>clean water</u>, <u>electricity</u> and <u>weekly rubbish collection</u>.

<u>Community involvement</u> is one of the most important parts of the project:

- <u>Residents choose which improvements</u> they want in their own favela, so they feel <u>involved</u>.
- <u>Neighbourhood associations</u> are formed to <u>communicate</u> with the residents and <u>make decisions</u>.
- The new <u>services</u> are <u>staffed</u> by <u>residents</u>, providing <u>income</u> and helping them to <u>learn new skills</u>.

> I'm bound to get a job with my new skills.

The Favela-Bairro Project has been Very Successful

1) The <u>standard of living</u> and <u>health</u> of residents have <u>improved</u>.

2) The <u>property values</u> in favelas that are part of the programme have increased by <u>80–120%</u>.

3) The number of <u>local businesses</u> within the favelas has almost <u>doubled</u>.

Her name is Rio and she dances on the recently widened and paved roads...

If you study a <u>different</u> squatter settlement redevelopment in class, and you'd rather write about that one in the exam then no problem — just make sure that you have <u>enough information</u> to cover the same <u>key points</u> on this page.

Urbanisation — Environmental Issues

It may be faster than a speeding bullet, but <u>rapid urbanisation</u> in <u>poorer countries</u> brings a whole host of <u>environmental problems</u>... which is a bit of a shame because that means you have to learn about them.

Rapid Urbanisation and Industrialisation Affect the Environment

<u>Rapid urbanisation</u> and <u>industrialisation</u> (where the <u>economy</u> of a country <u>changes</u> from being based on <u>agriculture</u> to <u>manufacturing</u>) can cause a number of <u>environmental problems</u>:

1) <u>Waste disposal problems</u> — people in cities <u>create a lot of waste</u>. This can <u>damage</u> people's <u>health</u> and the <u>environment</u>, especially if it's <u>toxic</u> and not <u>disposed of properly</u>.

2) <u>More air pollution</u> — this comes from <u>burning fuel</u>, vehicle <u>exhaust fumes</u> and <u>factories</u>.

3) <u>More water pollution</u> — water carries <u>pollutants from cities</u> into <u>rivers</u> and <u>streams</u>. For example, <u>sewage</u> and <u>toxic chemicals from industry</u> can get into rivers which causes serious <u>health problems</u>. <u>Wildlife</u> can also be <u>harmed</u>.

Waste Disposal is a Serious Problem in Poorer Countries

In richer countries, waste is disposed of by burying it in landfill sites, or by burning it. The amount of waste is also reduced by recycling schemes. <u>Poorer countries</u> struggle to dispose of the <u>large amount of waste</u> that's created by <u>rapid urbanisation</u> for many reasons:

1) <u>Money</u> — poorer countries often <u>can't afford</u> to dispose of waste <u>safely</u>, e.g. <u>toxic waste</u> has to be <u>treated</u> and this <u>can be expensive</u>. There are often <u>more urgent problems</u> to spend <u>limited funds</u> on, e.g. healthcare.

2) <u>Infrastructure</u> — poorer countries <u>don't</u> have the <u>infrastructure</u> needed, e.g. <u>poor roads</u> in <u>squatter settlements</u> mean <u>waste disposal lorries can't</u> get in to <u>collect rubbish</u>.

3) <u>Scale</u> — the problem is <u>huge</u>. A <u>large city</u> will generate <u>thousands of tonnes</u> of waste <u>every day</u>.

Air and Water Pollution Have Many Effects

Air pollution

Effects:

- Air pollution can lead to <u>acid rain</u>, which <u>damages buildings</u> and <u>vegetation</u>.

- It can cause <u>health problems</u> like <u>headaches</u> and <u>bronchitis</u>.

- Some pollutants <u>destroy</u> the <u>ozone layer</u>, which <u>protects us</u> from the sun's <u>harmful rays</u>.

Management of the pollution:

This can involve setting <u>air quality standards</u> for industries and constantly <u>monitoring levels</u> of pollutants to check they're <u>safe</u>.

Water pollution

Effects:

- Water pollution <u>kills fish</u> and other aquatic animals, which <u>disrupts food chains</u>.

- Harmful <u>chemicals</u> can <u>build up</u> in the <u>food chain</u> and <u>poison</u> humans who eat fish from the polluted water.

- Contamination of <u>water supplies</u> with <u>sewage</u> can <u>spread diseases</u> like <u>typhoid</u>.

Management of the pollution:

This can involve building <u>sewage treatment plants</u> and passing <u>laws</u> forcing factories to <u>remove pollutants</u> from their waste water.

Managing air and water pollution <u>costs a lot of money</u> and requires lots of different <u>resources</u>, e.g. <u>skilled workers</u> and <u>good infrastructure</u>. This makes it <u>harder</u> for poorer countries to <u>manage pollution</u>.

Waste disposal problems — what a load of rubbish...

The UK has <u>laws</u> that help to stop air and water pollution reaching <u>dangerous levels</u>, but <u>many poorer countries</u> have <u>no regulations</u> — you could pack a gas mask if you're visiting one, and a dry suit if you plan on having a dip...

Sustainable Cities

Cripes, it's the 's' word again — it wouldn't be a geography topic without it. This time we'll be discussing sustainable cities. Before starting, please put your eyelids in the open position and turn off your snoring...

Urban Areas Need to Become More Sustainable

1) Sustainable living means doing things in a way that lets the people living now have the things they need, but without reducing the ability of people in the future to meet their needs.

2) Basically, it means behaving in a way that doesn't irreversibly damage the environment or use up resources faster than they can be replaced.

3) For example, using only fossil fuels for power will add to climate change and eventually use them all up. This means the people in the future won't have any and the environment will be damaged — it's unsustainable.

4) Big cities need so many resources that it's unlikely they'd ever be truly sustainable. But things can be done to make a city (and the way people live there) more sustainable:

Schemes to reduce waste and safely dispose of it

More recycling means fewer resources are used, e.g. metal cans can be melted down and used to make more cans. Less waste is produced, which reduces the amount that goes to landfill. Landfill is unsustainable as it wastes resources that could be recycled and eventually there'll be nowhere left to bury the waste.

Safely disposing of toxic waste helps to prevent air and water pollution.

Conserving natural environments and historic buildings

Historic buildings, natural environments and open spaces are resources. If they get used up by people today (i.e. built on, or knocked down), they won't be available for people in the future to use. Historic buildings can be restored and natural environments can be protected. Existing areas of green space, like parks, should be left alone.

Building on brownfield sites

Brownfield sites are derelict areas that have been used, but aren't being used anymore. Using brownfield sites for new buildings stops green space being used up. So the space will still be available for people in the future. Developing brownfield sites also makes the city look nicer.

Building carbon-neutral homes

Carbon-neutral homes are buildings that generate as much energy as they use, e.g. by using solar panels to produce energy. For example, BedZED is a carbon-neutral housing development in London. More homes can be provided, without damaging the environment too much or causing much more pollution.

Creating an efficient public transport system

Good public transport systems mean fewer cars on the road, so pollution is reduced. Bus, train and tram systems that use less fuel and give out less pollution can also be used, e.g. some buses in London are powered by hydrogen and only emit water vapour.

5) People are much more likely to support sustainability initiatives like increased recycling or new public transport systems if they're involved in making the decisions about them. Including local people makes the schemes much more likely to succeed.

Conserve water — if it's brown, flush it down, if it's yellow, let it mellow...

Sustainability's a tough one to get your head around. Make sure you're clear on what it means before you memorise all the attempts to make it happen. And yes, toxic waste needs to be disposed of, not used to make fish with three eyes.

Sustainable Cities — Case Study

The <u>key points</u> you need to learn for this case study are: <u>where</u> the sustainable city is, its <u>size</u>, how the city is trying to be <u>sustainable</u>, what it <u>costs</u>, and how <u>successful</u> it is in being sustainable.

Curitiba is Aiming to be a Sustainable City

1) <u>Curitiba</u> is a city in southern <u>Brazil</u> with a population of <u>1.8 million people</u>.

2) The overall aims of its planners are to <u>improve</u> the <u>environment</u>, <u>reduce pollution</u> and <u>waste</u>, and <u>improve</u> the <u>quality of life</u> of residents.

3) The city has a <u>budget</u> of <u>$600 million</u> to spend <u>every year</u>.

4) Curitiba is working towards <u>sustainability</u> in different ways:

Brazil

Curitiba

---

1) <u>Reducing car use</u>

- There's a good <u>bus system</u>, used by more than <u>1.4 million passengers</u> per day.
- It's an '<u>express</u>' bus system — they have special <u>pre-pay boarding stations</u> that <u>reduce boarding times</u>, and <u>bus-only lanes</u> on the roads that <u>speed up journeys</u>.
- The same <u>cheap fare</u> is paid for all journeys, which <u>benefits poorer residents</u> who tend to live on the outskirts of the city.
- There are over <u>200 km</u> of <u>bike paths</u> in the city.
- The bus system and bike paths are so <u>popular</u> that <u>car use</u> is <u>25% lower</u> than the national average and Curitiba has one of the <u>lowest</u> levels of <u>air pollution</u> in Brazil.

---

2) <u>Plenty of open spaces and conserved natural environments</u>

- Green space <u>increased</u> from <u>0.5 m^2</u> per person in 1970 to <u>52 m^2</u> per person in 1990.
- It has over <u>1000 parks</u> and natural areas. Many of these were created in areas prone to <u>flooding</u>, so that the land is <u>useful</u> but <u>no serious damage</u> would be done if it flooded.
- Residents have planted <u>1.5 million trees</u> along the city's streets.
- Builders in Curitiba are given <u>tax breaks</u> if their building projects <u>include green space</u>.

---

3) <u>Good recycling schemes</u>

- <u>70%</u> of rubbish is <u>recycled</u>. Paper recycling saves the equivalent of <u>1200 trees per day</u>.
- Residents in poorer areas where the streets are too <u>narrow</u> for a weekly rubbish collection are given <u>food</u> and <u>bus tickets</u> for bringing their recycling in to local collection centres.

Curitiba has been Very Successful in its Aim to be Sustainable

1) The <u>reduction in car use</u> means that there's <u>less pollution</u> and <u>use of fossil fuels</u>. This means the environment <u>won't</u> be <u>damaged</u> so much for <u>people in the future</u>.

2) Leaving green, open spaces and conserving the natural environment means that <u>people in the future</u> will <u>still be able to use</u> the open spaces.

3) The high level of <u>recycling</u> means that <u>fewer resources</u> are used and <u>less waste</u> has to go to <u>landfill</u>. This means <u>more resources</u> will be available <u>in the future</u>.

4) Curitiba is also a <u>nice place to live</u> — <u>99%</u> of its residents said in a recent survey that they were happy with their town.

I'M FROM THE FUTURE.
I LIKE TO USE OPEN SPACES.

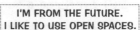

Oh Curitiba, it makes me happy...

I wish I lived In Curitiba. No traffic jams, buses that run on time, plenty of places to ride my bike and lots of lovely green parks. They even have a <u>flock of sheep</u> that goes around the parks to eat the grass instead of using lawnmowers.

Revision Summary for Changing Urban Environments

A refreshingly interesting topic that one. The basic gist is that more and more people are choosing to live in cities nowadays and this is causing some problems. In rich countries the problems are things like traffic jams and run-down CBDs, while in poorer countries things are more serious — people are forced to live in unsafe squatter settlements. So that's the general idea in this topic, but how much of the detail have you taken in? Time to find out...

1) Is most urbanisation happening today in rich countries or poor countries?

2) What is rural-urban migration?

3) Give one cause of rural-urban migration in poorer countries.

4) Give one cause of rural-urban migration in richer countries.

5) What does CBD stand for?

6) What's the main land use in the suburbs?

7) Explain what the rural-urban fringe is.

8) Give an example of how land use has changed in a part of a city.

9) Give two problems that cities in rich countries have to deal with.

10) Describe two solutions to one of these problems.

11) Give an example of a successful attempt to revitalise a CBD.

12) Name three problems caused by the increasing number of cars on the roads in richer countries.

13) Give one solution to reduce the number of cars in cities.

14) Give one reason why there is ethnic segregation in many urban areas.

15) Give one strategy used to support the multicultural mixture of urban areas.

16) What is a squatter settlement?

17) Describe what life is like for residents in a typical squatter settlement.

18) What is the informal sector of the economy?

19) a) Name a project that has improved a squatter settlement?

    b) Give one environmental improvement, one social improvement and one economic improvement that has happened.

    c) How successful has the project been?

20) Give one reason why waste disposal is such a problem in the cities of many poorer countries?

21) Give three effects of water pollution.

22) How can air pollution be managed?

23) What is meant by a sustainable city?

24) Explain how reducing waste can help a city to be more sustainable.

25) Explain how building on brownfield sites can help a city to become more sustainable.

26) a) Name a sustainable city.

    b) Describe three ways that the city is trying to become more sustainable.

    c) How successful do you think the city has been in its aim to be more sustainable? Give reasons for your answer.

Change in the Rural–Urban Fringe

Roll up, roll up, the <u>revision circus</u> is in <u>town</u>. Well, in the <u>rural-urban fringe</u>, anyway — the area right at the edge of a city, where there are <u>both urban land uses</u> (e.g. factories) and <u>rural land uses</u> (e.g. farming).

The Rural–Urban Fringe is a Popular Site for Development

As the <u>population</u> of an <u>urban area increases</u> the urban area <u>gets bigger</u>. This is called <u>urban sprawl</u>. One way it gets bigger is by <u>development</u> of the <u>rural-urban fringe</u>. Developments often include:

1) <u>Out-of-town retail outlets</u>.

2) <u>Leisure facilities</u>, e.g. golf courses, riding stables.

3) New <u>transport links</u>, e.g. new <u>motorways</u> connecting cities.

4) <u>Housing</u> — <u>more housing</u> is built in <u>existing villages</u>.

> **EXAMPLE:** <u>Golf courses</u>, <u>thousands</u> of <u>houses</u> and the <u>M5 motorway</u> have been built in the rural area <u>between</u> <u>Gloucester</u> and <u>Cheltenham</u>.

The rural-urban fringe is <u>popular for development</u> because:

1) There's <u>plenty of land available</u> and it's <u>cheaper</u> than in urban areas. Some developments are <u>huge</u> so can only be built where there's lots of land, e.g. <u>retail outlets</u> need <u>lots of space</u> for <u>car parking</u>.

2) It's <u>easy to reach</u> from the urban areas, e.g. people can <u>quickly drive</u> out to <u>retail outlets</u> or <u>golf courses</u> and there's plenty of room to <u>park</u>.

Fringe is soooo this season.

Development of the Rural–Urban Fringe has Impacts

1) <u>Traffic noise</u> and <u>pollution increase</u> as there's <u>more traffic</u>.

2) <u>People already living there</u> may feel the extra housing and developments <u>spoil the area</u>.

3) <u>Farmers</u> may be <u>forced to sell their land</u> so it can be <u>built on</u>, meaning they <u>can't earn a living</u>.

4) <u>Wildlife habitats</u> are <u>destroyed</u> by building on them.

Some urban areas have '<u>greenbelts</u>' around them though — a <u>ring of land</u> where <u>development is restricted</u>.

The Number and Size of Commuter Villages is Increasing

1) Some people <u>live in villages</u> and <u>commute</u> (travel) to work in <u>urban areas</u>. They choose to live there because it's a <u>nicer environment</u> and there's <u>less crime</u>, <u>pollution</u> and <u>noise</u> than in urban areas. Villages where there are <u>a lot of commuters</u> are called <u>commuter villages</u>.

2) <u>Transport</u> has become <u>cheaper</u> and <u>faster</u> in the 20th century. <u>Road and rail links</u> have <u>improved</u> too. This means <u>more people can live further away</u> from <u>where they work</u> and still commute to work easily. This has <u>increased</u> the <u>number</u> and <u>size</u> of <u>commuter villages</u>.

3) As a village becomes <u>more popular</u> it can cause <u>property prices</u> to <u>increase</u>. It can also cause an <u>increase</u> in <u>traffic congestion</u>.

Commuter villages are sometimes called suburbanised villages.

Growing Commuter Villages have Certain Characteristics

<u>Lots</u> of <u>services</u>, e.g. <u>shops</u>, <u>schools</u> and <u>restaurants</u>.

Lots of <u>middle-aged couples</u> with <u>children</u>, <u>professionals</u> and <u>wealthy</u> <u>retired people</u> who have moved there from the city as it's a <u>nicer environment</u>.

Lots of <u>new detached houses</u>, <u>converted barns</u> or <u>cottages</u> and <u>expensive estates</u>.

<u>Good public transport links</u>.

<u>Some jobs</u>, e.g. in local shops.

The soundtrack to the rural-urban fringe — a mix of house, garage and barn...

...a toe-tapping combination. But before you get carried away and start strutting your stuff, make sure you can reel off the <u>impacts</u> of <u>rural-urban fringe</u> development. Try and remember <u>at least three</u>, but <u>four</u> would be even better.

Change in Rural Areas — Case Study

Rural areas are changing, and it's not just that the country air is getting 'fresher' at muck-spreading time. Nope, there are economic and social changes too. Read on and you'll soon see exactly what I mean.

The Population of Some Rural Villages is Decreasing

There are two main reasons why:

1) Fewer jobs — the decline in agriculture and manufacturing in some rural areas means there are fewer jobs, so people have to move away to find work.

2) Growth in second home ownership — second homes are homes that people own as well as their main house. They usually use them at weekends or for holidays. The popularity of these properties increases house prices in the area so many young locals can't afford to live there and are forced to move away to somewhere they can afford a house.

A Decreasing Population Causes a Decrease in Services

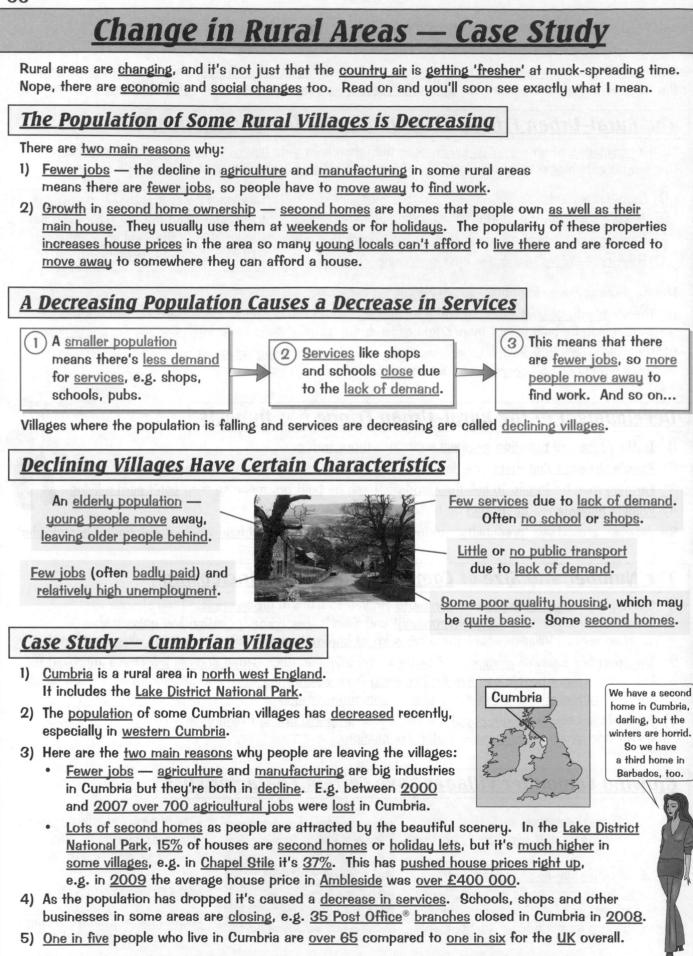

① A smaller population means there's less demand for services, e.g. shops, schools, pubs.

② Services like shops and schools close due to the lack of demand.

③ This means that there are fewer jobs, so more people move away to find work. And so on...

Villages where the population is falling and services are decreasing are called declining villages.

Declining Villages Have Certain Characteristics

An elderly population — young people move away, leaving older people behind.

Few jobs (often badly paid) and relatively high unemployment.

Few services due to lack of demand. Often no school or shops.

Little or no public transport due to lack of demand.

Some poor quality housing, which may be quite basic. Some second homes.

Case Study — Cumbrian Villages

1) Cumbria is a rural area in north west England. It includes the Lake District National Park.

2) The population of some Cumbrian villages has decreased recently, especially in western Cumbria.

3) Here are the two main reasons why people are leaving the villages:

- Fewer jobs — agriculture and manufacturing are big industries in Cumbria but they're both in decline. E.g. between 2000 and 2007 over 700 agricultural jobs were lost in Cumbria.

- Lots of second homes as people are attracted by the beautiful scenery. In the Lake District National Park, 15% of houses are second homes or holiday lets, but it's much higher in some villages, e.g. in Chapel Stile it's 37%. This has pushed house prices right up, e.g. in 2009 the average house price in Ambleside was over £400 000.

4) As the population has dropped it's caused a decrease in services. Schools, shops and other businesses in some areas are closing, e.g. 35 Post Office® branches closed in Cumbria in 2008.

5) One in five people who live in Cumbria are over 65 compared to one in six for the UK overall.

Cumbria

We have a second home in Cumbria, darling, but the winters are horrid. So we have a third home in Barbados, too.

In my village it's cool to lie down — it's got a reclining population...

At this point I reckon you should check you know why people leave some rural villages, and why this makes it harder to buy a pint or post a parcel. Make sure you've a couple of case study facts stashed away too, to impress the examiner.

Change in UK Commercial Farming

Commercial farming (farming to <u>make money</u>) in some rural areas <u>is changing</u>. Farmer Giles has been thrown off his tiny tractor by Slick Rick and his giant combine harvester. At least that's how I understand it.

Agri-business has Replaced Traditional Farming in Some Areas

1) In the UK, 60 years ago, there used to be lots of <u>small family farms</u> that sold a <u>mixture</u> of produce. Now there are lots of <u>large companies</u> that own <u>large farms</u>. They often produce a <u>single product</u>.

2) This kind of <u>large-scale commercial farming</u> is called <u>agri-business</u>.

3) <u>Modern farming practices</u> used by agri-businesses help to <u>maximise production and profits</u>. But the practices can take their toll on <u>the environment</u>:

- <u>Monoculture</u> (growing just <u>one</u> type of crop) <u>reduces biodiversity</u> as there are <u>fewer habitats</u>.
- <u>Removing hedgerows</u> to increase the area of farmland <u>destroys habitats</u>. It also <u>increases soil erosion</u> (hedgerows normally act as windbreaks).
- <u>Herbicides</u> are used to maximise crop production, but they can <u>kill wildflowers</u>.
- <u>Pesticides</u> are used to maximise crop production, but they can <u>kill other insects</u> as well as pests.
- <u>Fertilisers</u> are also used to maximise crop production, but they can <u>pollute rivers</u>, <u>killing fish</u>.
- <u>Making fertilisers</u>, <u>pesticides</u> and <u>herbicides</u> uses <u>fossil fuels</u>, which <u>adds</u> to <u>global warming</u>.

Biodiversity is the number and variety of organisms. A habitat is where an organism lives.

When fertilisers pollute rivers it's called eutrophication.

Organic Farming is Also Becoming More Common

1) <u>Organic farming</u> is basically farming <u>without</u> using <u>artificial pesticides</u> or <u>fertilisers</u>.

2) Methods used include <u>crop rotation</u> (changing the type of crop that's planted every year to stop pests building up), using <u>manure</u> as a fertiliser, <u>manual weeding</u> and using <u>biological control</u> (e.g. using ladybirds instead of pesticides to kill aphids). These methods can be <u>less damaging</u> to the <u>environment</u> than other methods.

3) Organic farming is <u>becoming more common</u>, e.g. in the UK in <u>1998</u>, <u>100 000 hectares</u> of land were organically farmed and in <u>2003</u> this had <u>increased</u> to <u>700 000 hectares</u>.

4) This is because the <u>demand for organic food</u> has <u>increased</u>. Some people buy organic food because they're <u>concerned</u> that modern farming practices <u>damage</u> the <u>environment</u> or that eating food that contains <u>pesticide</u> residues <u>might</u> be <u>harmful</u>.

Government Policies Aim to Reduce Farming's Environmental Impact

Here are two examples:

Environmental Stewardship Scheme

This involves <u>paying farmers money</u> for <u>every hectare</u> of land they <u>manage</u> in a way that <u>reduces</u> the <u>environmental impact</u>. For example, by <u>farming organically</u>.

Single Payment Scheme

This involves <u>paying farmers a subsidy</u> (paying them money to help them earn a decent living). But they're <u>only paid</u> it if they keep their land in a <u>good environmental condition</u>, e.g. if they leave 2 m around the edge of crop fields uncut to provide habitats. This <u>encourages</u> farmers to <u>reduce</u> the <u>environmental impact</u> of their farming.

Both schemes involve things like <u>using fewer chemicals</u> or leaving some areas <u>uncultivated</u>, so <u>less food</u> can be produced from the same area of land.
This can mean that <u>more land</u> has to be used for farming and the food produced is <u>more expensive</u>.

Monoculture is so 1990s — it's all about stereoculture these days...

It turns out <u>farming</u> isn't as simple as throwing a few seeds around a field and making the world's best scarecrow. Nope it's a serious business. <u>Agri-business</u>, in fact, so make sure you know the definition.

Change in UK Commercial Farming — Case Study

More exciting info on UK farming and a nice case study to get your teeth into. No need to thank me.

The Prices Farmers can Charge May be Decided by Supermarkets

1) Four major supermarket chains now control 75% of grocery sales in the UK.

2) This means farmers often have no choice but to cut their prices when asked to by the supermarkets, as there's no-one else to sell to. If they don't cut their prices the supermarkets will find other suppliers.

3) Many foods need processing before supermarkets will buy them, so sometimes farm products are bought by a processing firm. That firm then sells the finished product on to the supermarket at a profit. This adds another step to the supply chain, which can further reduce prices paid to farmers.

4) Some farmers struggle to earn a living due to the low prices (and some go out of business).

UK Farmers Now Have to Compete with the Global Market

1) Before the 1960s most of the food people ate was grown in the UK, usually in the local area.

2) Since then there's been an increase in the global trading of food with more and more of our food being imported from other countries.

3) This has helped to provide enough food for the growing population and has meant people in the UK can get a wide range of food all year round.

4) Imported food is often cheaper if it's grown in poorer countries where farmers pay less for land and pay less for workers to harvest it. UK farmers have to compete with these lower prices.

(Dramatization)

Transporting food a long way produces lots of CO_2, which adds to global warming.

Case Study — East Anglia

1) East Anglia is an area that includes Norfolk, Suffolk, Cambridgeshire and Essex.

2) It's known as the UK's 'bread basket' because it produces more than a quarter of England's wheat and barley. Farms in the area also produce 2.2 million eggs every day.

3) Agri-business has increased in East Anglia, e.g. in Essex the number of farms over 100 hectares increased from 828 in 1990 to 849 in 2005.

East Anglia

4) Organic farming in East Anglia has increased, but it's still quite low in the area, e.g. in 2008 1.3% of farmland was farmed organically, compared to 3.7% for England overall.

5) Farmers in East Anglia are trying to reduce the environmental impact of farming — the area has more land covered by the Environmental Stewardship Scheme (see previous page) than any other area in the UK.

6) Farmers in East Anglia have been affected by supermarket prices and competition from overseas. E.g. in 1997 peas from East Anglia sold at 25p per kilo but by 2002 this had dropped to 17p.

East Anglia — the home of bread and trigonometry...

It's fine if you've learnt a different case study in class and want to use that instead. You just need to understand all the different changes to commercial farming in that area and what's being done about the impacts of farming there too.

Sustainable Rural Living

Ah, the countryside. It's all about running through the poppy fields flicking your shiny, newly-shampooed hair. It feels like it could go on forever. Perhaps it can, but only if we run and flick in a sustainable way.

Rural Living Needs to be Sustainable

Sustainable living means living in a way that lets the people alive now get the things they need, but without stopping people in the future getting the things they need. Basically it means behaving in a way that doesn't irreversibly damage the environment or use up resources faster than they can be replaced. There are two main reasons why rural living can be unsustainable:

1) High car use — many rural areas have little public transport (often due to low demand) so lots of people travel by car. This uses up fossil fuel resources and releases carbon dioxide, which adds to global warming.

2) Use of some farming techniques:
   - Some farming techniques use up fossil fuels. For example, some farms use a lot of artificial fertilisers — making artificial fertilisers uses up fossil fuel.
   - Irrigation of farmland can also deplete water resources.
   - Some techniques damage the environment (see page 97 for a list).

If I can't use pesticides...

There are ways to make rural living more sustainable:

1) Conserve resources such as water and fossil fuels. For example, by:
   - Using public transport more to reduce car use.
   - Using irrigation techniques that don't waste water, e.g. drip-irrigation.

Drip-irrigation uses pipes to deliver drops of water directly to plant roots.

2) Protect the environment. For example, by:
   - Reducing the use of herbicides, fertilisers and pesticides to reduce their impacts.
   - Maintaining hedgerows to provide wildlife habitats.

Any attempts to make rural living sustainable have to support the needs of the rural population, e.g. you can't just stop all farming to reduce its environmental impacts because local people need a source of income.

Government Initiatives Protect the Rural Economy and Environment

There are various government initiatives (schemes) to help protect the rural environment and the rural economy — they help towards sustainable living. Here are a couple of examples:

1 Community Rail Partnerships help increase local train use by improving bus links, developing cycle routes to stations and improving station buildings. This reduces car use and the environmental impacts it has.

2 The Rural Development Programme for England gives farmers financial support to diversify their farms, e.g. to provide bed and breakfast accommodation or set up a tourist attraction. This gives farmers an extra income, so they're not as dependent on farming. It can also reduce the environmental impact of farming as some farmers don't need to farm as much.

3 The Environmental Stewardship Scheme involves paying farmers money to manage their land in a way that reduces the environmental impact (see p. 97).

Rural living — it's not all tweed jackets, tractors and chewing straw...

First get your head around why rural living can be unsustainable, then even if you can't remember how to make it more sustainable in the exam, you should be able to work it out. Another pearl of wisdom... sometimes I think I'm too nice.

Changes to Farming in Tropical Areas

I love bananas in custard, but definitely not in crumble. This <u>tropical farming</u> page isn't about them though.

Subsistence Farming is being Replaced by Commercial Farming

1) <u>Subsistence farming</u> is where farmers only <u>produce enough food</u> to <u>feed their families</u>. In <u>tropical areas</u> farmers usually <u>clear</u> an area of <u>rainforest</u> to make land for producing food. The <u>soil</u> quickly becomes <u>infertile</u> though, so the farmers <u>move to another area</u> and <u>start again</u>. This is called <u>shifting cultivation</u>.

2) Subsistence farming is being <u>replaced</u> in some tropical areas by <u>commercial farming</u> — where crops and animals are produced to be <u>sold</u>, e.g. <u>coffee</u>, <u>cotton</u>, <u>sugar cane</u> and <u>cattle</u>.

3) Sometimes <u>subsistence farmers switch</u> to commercial farming, and sometimes <u>big companies set up farms</u> (they often <u>take over</u> subsistence farmers' land).

4) Commercial farm products are usually sold to <u>richer countries</u>.

5) This has a few <u>impacts</u>:
   - <u>Subsistence farmers</u> who've had their land taken by big companies are forced onto <u>poorer land</u> where it's <u>harder</u> to <u>grow food for themselves</u>.
   - If farmers are <u>dependent</u> on a <u>single crop or animal</u> and <u>prices drop</u> they might <u>not</u> have <u>enough money</u> to <u>buy food</u>. It also means farmers will only have an <u>income</u> around <u>harvest or slaughter time</u> — if they can't make a lot of money, they'll <u>struggle</u> to <u>buy food</u> for the <u>rest of the year</u>.
   - There <u>isn't as much food being produced locally</u>, so food has to be <u>brought in</u> from <u>further away</u>. This <u>increases food prices</u>.

The best a farmer can get.

Commercial farming in tropical areas can also be called cash cultivation.

Irrigation has Changed Agriculture

Growing <u>lots of crops</u> to sell <u>needs a lot of water</u>, so farmers often have to <u>irrigate</u> their land (<u>artificially apply water</u>). Irrigation has <u>physical</u> and <u>human</u> impacts:

	Positive	Negative
Physical impacts	• <u>More land</u> can be farmed. • <u>Crop yields</u> are <u>higher</u> and <u>fewer harvests</u> are <u>lost</u> due to lack of water. • <u>High yields</u> mean farmers <u>don't</u> need to <u>clear more land</u> for farming, e.g. by <u>deforestation</u>.	• Irrigation can cause <u>soil erosion</u>. • Without proper drainage <u>salt</u> can <u>build up</u> (<u>salinisation</u>) causing <u>crops to fail</u>. • If the land <u>isn't well drained</u> it becomes <u>waterlogged</u> so <u>nothing can grow</u>.
Human impacts	• <u>Higher yields</u> mean <u>more food</u>. This <u>decreases</u> the risk of <u>famine</u>. • Higher yields mean farmers make <u>more profit</u> — giving them a <u>better quality of life</u>.	• Large-scale irrigation projects can be <u>expensive</u> and cause <u>rural debt</u> to increase. • <u>Mosquitoes</u> that <u>spread malaria</u> breed in <u>irrigation ditches</u>. • <u>Waterborne diseases</u> can also become <u>more common</u>.

Appropriate Technologies have also Changed Agriculture

<u>Appropriate technologies</u> are <u>simple</u>, <u>low cost</u> technologies that <u>increase food production</u>. They're <u>made</u> and <u>maintained</u> using <u>local knowledge</u> and <u>resources</u>, so they're not <u>dependent</u> on any <u>outside support</u>, <u>expensive equipment</u> or <u>fuel</u>. Here are two examples:

1) The <u>treadle pump</u> is a <u>human-powered</u> pump used in <u>Bangladesh</u>. It <u>pumps water</u> from below the ground to <u>irrigate small areas</u> of <u>land</u>. This is important in Bangladesh as the main crop (<u>rice</u>) needs <u>lots</u> of <u>water</u> to grow. It costs US $7 to buy and it's <u>increased</u> Bangladeshi farmers' average annual <u>incomes</u> by roughly $100 because of increased crop yields.

2) <u>Lines of stones</u> are used to <u>trap water</u> on <u>sloping fields</u> in <u>Burkina Faso</u>. It <u>increases</u> the amount of <u>water</u> that <u>soaks into</u> the <u>soil</u> so <u>more</u> is <u>available</u> for crops. It's <u>increased crop yields</u> by about <u>50%</u>.

The fanciest option isn't always the best, as many WAG weddings have proved...

The idea of <u>appropriate technology</u> is pretty important. You could be asked whether an <u>irrigation system</u> would be appropriate technology for a tropical region. Think about how it's <u>suited</u> to the area and what <u>problems</u> it could cause.

Factors Affecting Farming in Tropical Areas

Just a second, I hope I didn't just see your eyes glazing over. No, I didn't think I did. Don't worry, there are just a few more <u>tropical farming issues</u> to learn about and then this topic is all finished.

Soil Erosion can be a Big Problem for Tropical Farmers

<u>Soil erosion</u> happens naturally due to the action of <u>wind</u> and <u>rain</u>. Soil erosion is <u>common</u> in <u>tropical areas</u> because there's <u>heavy rainfall</u>, which <u>washes away</u> the soil. <u>Overgrazing</u> can <u>cause erosion</u> because <u>plants</u> that <u>hold the soil together</u> are <u>removed</u>. Soil erosion can cause <u>serious problems</u>:

1) Erosion of the <u>nutrient-rich top layer</u> of soil makes the soil <u>unsuitable for farming</u> — it <u>doesn't</u> have <u>enough nutrients</u> and it <u>can't hold water as well</u>.

2) When the <u>land can't be farmed anymore</u>, the farmers either have to <u>move away</u> (e.g. to urban areas, see below) or they have to <u>clear more land</u> and start again.

3) The eroded soil is <u>washed into rivers</u>, which <u>raises riverbeds</u>. This means the rivers <u>can't hold as much water</u> and are <u>more likely</u> to <u>flood</u>.

Mining and Forestry Affect Subsistence Farming

A lot of <u>mining</u> and <u>forestry</u> goes on in <u>tropical rainforests</u>, which <u>affects subsistence farming</u> there:

MINING

1) Mining <u>companies</u> can <u>force local people off</u> their <u>land</u>. This means the local farmers have <u>no source of income</u>.

2) Mining <u>uses lots of water</u>. This can <u>reduce crop yields</u> for local farmers because there's <u>less water</u> for <u>irrigation</u>.

3) After the resources have all been <u>extracted</u> the <u>land</u> is often left <u>unusable</u> (e.g. because of <u>pollution</u>). This means there's <u>less land available</u> for local farmers.

FORESTRY

1) <u>Deforestation</u> can make <u>floods more common</u> as there are <u>fewer trees</u> to <u>intercept rainfall</u>. Floods can <u>waterlog soil</u>, <u>reducing crop yields</u>. They can also <u>wash away crops</u>.

2) Without trees, <u>less water</u> is <u>removed</u> from the <u>soil</u> and <u>evaporated</u> into the atmosphere. This means <u>fewer clouds form</u> and <u>rainfall</u> in the area is <u>reduced</u>. Reduced rainfall means <u>lower crop yields</u> for local farmers.

3) It's not all bad though — deforestation means <u>more land</u> is <u>available</u> for <u>farming</u>, so farmers can <u>increase their income</u>.

Farming Difficulties Lead to Rural-Urban Migration

1) Factors such as <u>soil erosion</u>, <u>mining</u> and <u>forestry</u> can cause <u>farms</u> to <u>fail</u>.

2) This means farmers <u>can't</u> make a <u>profit</u> or <u>grow enough</u> to <u>feed themselves</u> and <u>their families</u>.

3) People are <u>forced</u> to <u>abandon their land</u> and look for <u>other work</u>. They <u>leave</u> the <u>countryside</u> and <u>move</u> to <u>towns</u> and <u>cities</u> (this is known as <u>rural-urban migration</u>).

4) However, there <u>aren't enough jobs</u> or <u>houses</u> in the cities for <u>all</u> the people that move there. This means things like <u>squatter settlements</u> spring up (poor quality houses <u>built illegally</u>).

5) As more land is abandoned, <u>less food</u> is <u>produced</u> by the country. This causes <u>food prices</u> to <u>rise</u> due to the <u>cost</u> of importing it from <u>other countries</u>.

6) Governments can <u>reduce</u> rural-urban migration by <u>helping farmers</u>, e.g. by encouraging the use of <u>appropriate technology</u> to decrease <u>water shortages</u> and <u>educating farmers</u> about <u>sustainable</u> farming methods.

Abandon land!

What a lot of cow pats — I blame failing farms...

More <u>impacts</u> to learn again here, but I guess you're getting used to that by now. If not, I recommend a smidge more revision. Cover the page and write down as many <u>impacts</u> of <u>mining</u> and <u>forestry</u> on <u>subsistence farming</u> as you can.

Revision Summary for Changing Rural Environments

At last, the end of another long, hard topic — well nearly the end. Before you stop for a well-earned cup of tea and an episode of Hollyoaks, there's just the small matter of this list of questions.
It's really in your best interests to have a look through them, because if there are any you can't answer then you can bet your favourite pair of pants that that's exactly what will come up in the exam.

1) List four types of development often found in the rural-urban fringe.

2) Give two reasons why the rural-urban fringe is a popular place for these developments.

3) Give four impacts of development in the rural-urban fringe.

4) Give two reasons why people live in commuter villages.

5) List four characteristics of commuter villages.

6) What are the two main reasons why the populations of some rural villages are decreasing?

7) Explain why a decrease in population in a village can cause a decrease in services.

8) Give four characteristics of declining villages.

9) a) Give an example of a rural area in the UK with a declining population.

   b) Give one cause of depopulation in that area.

   c) Give one problem resulting from depopulation in that area.

10) Define the term agri-business.

11) What is monoculture?

12) Give two ways modern farming practices can affect the environment.

13) Explain what organic farming is.

14) Describe one government policy aimed at reducing the environmental impact of farming.

15) Explain how supermarkets may influence the prices farmers charge.

16) How does competition from the global market affect the prices UK farmers can get for their produce?

17) a) Give an example of a commercial farming area in the UK.

    b) Give two ways farming in the area has changed recently.

    c) Give one way the environmental impact of farming in the area has been reduced.

18) What is meant by sustainable living?

19) Give two ways that rural living can be unsustainable.

20) Describe two ways rural living can be made more sustainable.

21) Describe three government initiatives that are designed to boost the economy or protect the environment in rural areas.

22) Explain what is meant by:

    a) Subsistence farming.

    b) Commercial farming.

23) Give three problems switching from subsistence farming to commercial farming can cause.

24) Give two physical and two human impacts of irrigation.

25) What is meant by appropriate technology?

26) Explain why soil erosion is a problem for tropical farmers.

27) How can mining in an area have a negative impact on subsistence farming?

28) How can forestry in an area have a negative impact on subsistence farming?

29) Explain how factors such as soil erosion, mining and forestry can lead to rural-urban migration by farmers in tropical countries.

Measuring Development

OK, this topic's a <u>bit trickier</u> than the other human ones, but <u>fear not</u> — I'll take it slowly.

Development is when a Country is Improving

When a country <u>develops</u> it basically <u>gets better</u> for the people living there — their <u>quality of life improves</u> (e.g. their <u>wealth</u>, <u>health</u> and <u>safety</u>). The level of development is different in <u>different countries</u>, e.g. France is more developed than Ethiopia.

There Are Loads of Measures of Development

Development is <u>pretty hard to measure</u> because it <u>includes so many things</u>. But you can <u>compare</u> the development of different countries using '<u>measures of development</u>'. You need to <u>know</u> these ones:

NAME	WHAT IT IS	A MEASURE OF...	AS A COUNTRY DEVELOPS IT GETS...
Gross Domestic Product (GDP)	The <u>total value</u> of <u>goods</u> and <u>services</u> a <u>country produces</u> in a <u>year</u>. It's often given in US dollars (US$).	Wealth	Higher ↑
Gross National Income (GNI)	The <u>total value</u> of <u>goods</u> and <u>services people of that nationality produce</u> in a <u>year</u> (i.e. GDP + money from people living abroad). It's often given in US$. It's also called Gross National Product (GNP).	Wealth	Higher ↑
GNI per head	This is the <u>GNI divided by</u> the <u>population</u> of a country. It's sometimes called GNI per capita.	Wealth	Higher ↑
Birth rate	The number of <u>live babies born per thousand</u> of the population <u>per year</u>.	Womens' rights	Lower ↓
Death rate	The number of <u>deaths per thousand</u> of the population <u>per year</u>.	Health	Lower ↓
Infant mortality rate	The number of <u>babies</u> who <u>die under 1 year old</u>, <u>per thousand babies born</u>.	Health	Lower ↓
People per doctor	The <u>average number</u> of people <u>for each doctor</u>.	Health	Lower ↓
Literacy rate	The <u>percentage</u> of <u>adults</u> who can <u>read and write</u>.	Education	Higher ↑
Access to safe water	The <u>percentage</u> of people who can <u>get clean drinking water</u>.	Health	Higher ↑
Life expectancy	The <u>average age</u> a person can <u>expect to live to</u>.	Health	Higher ↑
Human Development Index (HDI)	This is a number that's calculated using <u>life expectancy</u>, <u>literacy rate</u>, <u>education level</u> (e.g. degree) and <u>income per head</u>.	Lots of things	Higher ↑

Many of these measures are <u>linked</u> — there's a <u>relationship between them</u> (the posh name for this is a <u>correlation</u>). For example, countries with <u>high GNI</u> tend to have <u>low death rates</u> and <u>high life expectancy</u> because they have <u>more money</u> to <u>spend on healthcare</u>. Countries where a high percentage of people have <u>access to clean water</u> have <u>low infant mortality rates</u> because <u>fewer babies die</u> from <u>waterborne diseases</u>.

These Measures Have Limitations When Used On Their Own

1) The measures can be <u>misleading</u> when used <u>on their own</u> because they're <u>averages</u> — they <u>don't show up elite groups</u> in the population or <u>variations within the country</u>. For example, if you looked at the GNI of Iran it might seem quite developed (because the GNI is quite high), but in reality there are some really wealthy people and some poor people.

2) They also shouldn't be used on their own because as a country develops, some aspects <u>develop before others</u>. So it might seem that a country's <u>more developed</u> than it <u>actually is</u>.

3) Using <u>more than one measure</u> or using the <u>HDI</u> (which uses lots of measures) <u>avoids these problems</u>.

Measures of revision — they're called exams...

...and talking of exams you'd better get <u>learnin'</u> the <u>11 measures of development</u> listed above. Make sure you know what <u>each one means</u> and whether it gets <u>higher</u> or <u>lower</u> as a <u>country develops</u>. In fact, shut the book and test yourself now.

Global Inequalities

'Global inequalities' means the level of <u>development</u> of <u>different countries</u> in the world is <u>unequal</u>.

Some Countries are More Developed than Others

1) Countries used to be classified into <u>two</u> categories based on <u>how economically developed</u> they were.

2) <u>Richer</u> countries were classed as <u>More Economically Developed Countries</u> (MEDCs) and <u>poorer</u> countries were classed as <u>Less Economically Developed Countries</u> (LEDCs).

3) <u>MEDCs</u> were generally found in the <u>north</u>. They included the USA, European countries, Australia and New Zealand.

4) <u>LEDCs</u> were generally found in the <u>south</u>. They included India, China, Mexico, Brazil and all the African countries.

5) But using this simple classification you <u>couldn't tell</u> which countries were <u>developing quickly</u> and which <u>weren't really developing at all</u>. Nowadays, countries are classified into <u>more categories</u>, for example:

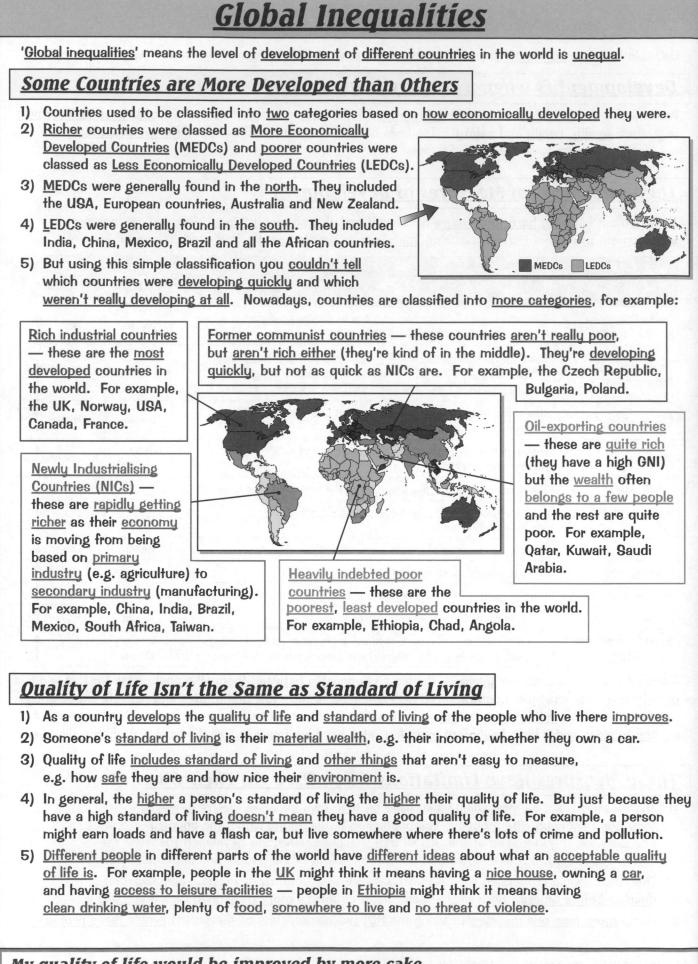

| ■ MEDCs | □ LEDCs |

<u>Rich industrial countries</u> — these are the <u>most developed</u> countries in the world. For example, the UK, Norway, USA, Canada, France.

<u>Former communist countries</u> — these countries <u>aren't really poor</u>, but <u>aren't rich either</u> (they're kind of in the middle). They're <u>developing quickly</u>, but not as quick as NICs are. For example, the Czech Republic, Bulgaria, Poland.

<u>Newly Industrialising Countries (NICs)</u> — these are <u>rapidly getting richer</u> as their <u>economy</u> is moving from being based on <u>primary industry</u> (e.g. agriculture) to <u>secondary industry</u> (manufacturing). For example, China, India, Brazil, Mexico, South Africa, Taiwan.

<u>Oil-exporting countries</u> — these are <u>quite rich</u> (they have a high GNI) but the <u>wealth</u> often <u>belongs to a few people</u> and the rest are quite poor. For example, Qatar, Kuwait, Saudi Arabia.

<u>Heavily indebted poor countries</u> — these are the <u>poorest</u>, <u>least developed</u> countries in the world. For example, Ethiopia, Chad, Angola.

Quality of Life Isn't the Same as Standard of Living

1) As a country <u>develops</u> the <u>quality of life</u> and <u>standard of living</u> of the people who live there <u>improves</u>.

2) Someone's <u>standard of living</u> is their <u>material wealth</u>, e.g. their income, whether they own a car.

3) Quality of life <u>includes standard of living</u> and <u>other things</u> that aren't easy to measure, e.g. how <u>safe</u> they are and how nice their <u>environment</u> is.

4) In general, the <u>higher</u> a person's standard of living the <u>higher</u> their quality of life. But just because they have a high standard of living <u>doesn't mean</u> they have a good quality of life. For example, a person might earn loads and have a flash car, but live somewhere where there's lots of crime and pollution.

5) <u>Different people</u> in different parts of the world have <u>different ideas</u> about what an <u>acceptable quality of life is</u>. For example, people in the <u>UK</u> might think it means having a <u>nice house</u>, owning a <u>car</u>, and having <u>access to leisure facilities</u> — people in <u>Ethiopia</u> might think it means having <u>clean drinking water</u>, plenty of <u>food</u>, <u>somewhere to live</u> and <u>no threat of violence</u>.

My quality of life would be improved by more cake...

Unfortunately, revising development categories is a <u>teeny bit more difficult</u> now that there are a <u>few categories</u> rather than just two (I wish those geographers would make their minds up about what to call things — sooooo inconsiderate).

Causes of Global Inequalities

Zzzzzz... oh, sorry, I nodded off for a minute there. You need to know the reasons why there are global inequalities — i.e. why countries differ in how developed they are. There are a fair few, but you'll be OK...

Environmental Factors Affect How Developed a Country Is

A country is more likely to be less developed if it has...

1 A POOR CLIMATE

1) If a country has a poor climate (really hot or really cold) they won't be able to grow much.
2) This reduces the amount of food produced.
3) In some countries this can lead to malnutrition, e.g. in Chad and Ethiopia.
   People who are malnourished have a low quality of life.
4) People also have fewer crops to sell, so less money to
   spend on goods and services. This also reduces their quality of life.
5) The government gets less money from taxes (as less is sold and bought).
   This means there's less to spend on developing the country,
   e.g. to spend on improving healthcare and education.

2 POOR FARMING LAND

If the land in a country is steep or has poor soil
(or no soil) then they won't produce a lot of food.
This has the same effect as a poor climate (see above).

3 LIMITED WATER SUPPLIES

Some countries don't have a lot of water,
e.g. Egypt, Jordan. This makes it harder for
them to produce a lot of food. This has the
same effect as a poor climate (see above).

4 LOTS of NATURAL HAZARDS

1) A natural hazard is an event that has the potential to affect people's lives or property,
   e.g. earthquakes, tsunamis, volcanic eruptions, tropical storms, droughts, floods.
2) When natural hazards do affect people's lives or property they're called natural disasters.
3) Countries that have a lot of natural disasters have to spend a lot of money rebuilding
   after disasters occur, e.g. Bangladesh.
4) So natural disasters reduce quality of life for the people affected, and they reduce
   the amount of money the government has to spend on development projects.

5 FEW RAW MATERIALS

1) Countries without many raw materials like coal, oil or metal ores tend to make less money
   because they've got fewer products to sell.
2) This means they have less money to spend on development.
3) Some countries do have a lot of raw materials but still aren't very developed because they
   don't have the money to develop the infrastructure to exploit them (e.g. roads and ports).

There are Three Main Political Factors that Slow Development

1) If a country has an unstable government it might not invest in things like healthcare, education
   and improving the economy. This leads to slow development (or no development at all).
2) Some governments are corrupt. This means that some people in the country get richer
   (by breaking the law) while the others stay poor and have a low quality of life.
3) If there's war in a country the country loses money that could be spent on development —
   equipment is expensive, buildings get destroyed and fewer people work (because they're fighting).
   War also directly reduces the quality of life of the people in the country.

Hot and dry — good for holidays, bad for development...

Basically, if a country is rubbish for farming, tropical storms keep wrecking the place or it has a dodgy government, then it's going to struggle to develop. There are a few exceptions though, e.g. Japan gets battered by natural hazards but is developed.

Causes of Global Inequalities

Countries really do have a tough time trying to develop. It's not just things like earthquakes and a shortage of water that hold them back — things like trade, debt and mucky water are to blame too...

Economic Factors Affecting Development Include Trade and Debt

A country is more likely to be less developed if it has...

1 POOR TRADE LINKS

1) Trade is the exchange of goods and services between countries.

2) World trade patterns (who trades with who) seriously influence a country's economy and so affect their level of development.

3) If a country has poor trade links (it trades a small amount with only a few countries) it won't make a lot of money, so there'll be less to spend on development.

2 LOTS of DEBT

1) Very poor countries borrow money from other countries and international organisations, e.g. to help cope with the aftermath of a natural disaster.

2) This money has to be paid back (sometimes with interest).

3) Any money a country makes is used to pay back the money, so isn't used to develop.

3 AN ECONOMY BASED ON PRIMARY PRODUCTS

1) Countries that mostly export primary products (raw materials like wood, metal and stone) tend to be less developed.

2) This is because you don't make much profit by selling primary products. Their prices also fluctuate — sometimes the price falls below the cost of production.

3) This means people don't make much money, so the government has less to spend on development.

4) Countries that export manufactured goods tend to be more developed.

5) This is because you usually make a decent profit by selling manufactured goods. Wealthy countries can also force down the price of raw materials that they buy from poorer countries.

Social Factors Affect Development Too

1 DRINKING WATER

1) A country will be more developed if it has clean drinking water available.

2) If the only water people can drink is dirty then they'll get ill — waterborne diseases include typhoid and cholera. Being ill a lot reduces a person's quality of life.

3) Ill people can't work, so they don't add money to the economy, and they also cost money to treat.

4) So if a country has unsafe drinking water they'll have more ill people and so less money to develop.

2 THE PLACE OF WOMEN IN SOCIETY

1) A country will be more developed if women have an equal place with men in society.

2) Women who have an equal place in society are more likely to be educated and to work.

3) Women who are educated and work have a better quality of life, and the country has more money to spend on development because there are more people contributing to the economy.

3 CHILD EDUCATION

1) The more children that go to school (rather than work) the more developed a country will be.

2) This is because they'll get a better education and so will get better jobs. Being educated and having a good job improves the person's quality of life and increases the money the country has to spend on development.

Debt — the cause of slow development and my lack of a fancy mobile phone...

Reading this page makes you realise how nice the UK is — there's trade galore, clean water and women work. Don't forget there are always exceptions to the rules above, e.g. countries that export oil (a primary product) are often quite rich.

Global Inequalities — Case Study

Thought you could get away without learning a <u>case study</u> for why there are <u>global inequalities</u> (why some countries are <u>less developed</u> than others), think again buster...

Hurricane Mitch Hit Nicaragua and Honduras in October 1998

Hurricane Mitch hit a few countries in <u>1998</u>, but <u>Nicaragua</u> and <u>Honduras</u> were the <u>worst hit</u>. Here are some of the <u>impacts</u> in each country:

Nicaragua

1) Around <u>3000</u> people were <u>killed</u>.
2) The impact on <u>agriculture</u> was high — <u>crops failed</u> and <u>50 000 animals died</u>.
3) <u>70%</u> of <u>roads</u> were <u>unusable</u> and <u>71 bridges</u> were <u>damaged</u> or <u>destroyed</u>.
4) <u>23 900 houses</u> were <u>destroyed</u> and <u>17 600</u> more were <u>damaged</u>.
5) <u>340 schools</u> and <u>90 health centres</u> were <u>damaged</u> or <u>destroyed</u>.

Honduras

1) Around <u>7000</u> people were <u>killed</u>.
2) The hurricane <u>destroyed 70%</u> of the country's <u>crops</u>, e.g. bananas, rice, coffee beans.
3) Around <u>70-80%</u> of the <u>transport infrastructure</u> (e.g. roads and bridges) was <u>severely damaged</u>.
4) <u>35 000 houses</u> were <u>destroyed</u> and <u>50 000</u> more were <u>damaged</u>.
5) <u>20% of schools</u> were <u>damaged</u>, as well as <u>117 health centres</u> and <u>six hospitals</u>.

Hurricane Mitch Set Back Development in Nicaragua...

1) In 1998 the <u>GDP grew</u> by <u>4%</u>, which was <u>less than estimated</u>. The rate of growth <u>slowed</u> in the <u>later months</u> of 1998 — <u>after</u> Hurricane Mitch hit.
2) <u>Exports</u> of <u>rice</u> and <u>corn went down</u> because <u>crops</u> were <u>damaged</u> by the hurricane. This meant <u>people earnt less money</u>, so were poorer, and the <u>government</u> had <u>less to spend</u> on development.
3) The <u>total damage</u> caused by the hurricane is estimated to be <u>$1.2 billion</u>. The cost of <u>repairs</u> took <u>money away</u> from <u>development</u>.
4) The <u>education</u> of <u>children suffered</u> — the number of <u>children</u> that <u>worked</u> (rather than went to school) <u>increased</u> by <u>8.1%</u> after the hurricane. This meant the children had a <u>lower quality of life</u> and found it <u>harder</u> to get <u>good jobs</u> later in life.

...and In Honduras

1) In 1998 <u>money from agriculture</u> made up <u>27%</u> of the country's <u>GDP</u>. In <u>2000</u> this had <u>fallen</u> to <u>18%</u> because of the <u>damage to crops</u> caused by the hurricane. This <u>reduced</u> the <u>quality of life</u> for people who worked in <u>agriculture</u> because they made <u>less money</u>.
2) <u>GDP</u> was <u>estimated</u> to <u>grow 5%</u> in 1998, but it <u>only grew 3%</u> due to the hurricane. This meant there was <u>less money available</u> for <u>development</u> than there would have been if Mitch hadn't hit.
3) The cost of <u>repairing</u> and <u>rebuilding houses</u>, <u>schools</u> and <u>hospitals</u> was estimated to be <u>$439 million</u> — this money <u>could</u> have been <u>used to develop the country</u>.
4) All these things <u>set back development</u> — the <u>Honduran President</u> claimed the hurricane <u>destroyed 50 years of progress</u>.

Nicaragua is an anagram of iguana car — brum brum...

There are tons of facts on this page for you to cram into your brain. If you already <u>know a case study</u> of how a natural disaster affected development that's fine, but if not, <u>shut the book</u> and <u>scribble what you can remember</u> till you get it <u>all</u>.

Reducing Global Inequality

As you might have gathered, global inequality is on the list of bad things in the world (along with revision, exams and broccoli). So some people are trying to <u>reduce it</u> by <u>helping poorer countries develop</u>.

Some People are Trying to Improve Their Own Quality of Life

Some people in poorer countries try to <u>improve</u> their <u>quality of life</u> on their <u>own</u> — rather than relying on <u>help from others</u>. This is called '<u>self-help</u>'. Here are a few ways that people do this:

1) <u>Moving</u> from <u>rural</u> areas to <u>urban</u> areas often improves a person's quality of life. Things like <u>water</u>, <u>food</u> and <u>jobs</u> are often <u>easier to get</u> in towns and cities.

2) Some people improve their quality of life by <u>improving</u> their <u>environment</u>, e.g. their <u>houses</u>.

3) <u>Communities</u> can <u>work together</u> to improve quality of life for everyone in the community, e.g. some communities <u>build</u> and <u>run services</u> like <u>schools</u>.

Fair Trade and Trading Groups Help Increase the Money Made from Trade

Fair trade

1) Fair trade is all about getting a <u>fair price</u> for <u>goods produced</u> in <u>poorer countries</u>, e.g. coffee.

2) Companies who want to <u>sell products</u> labelled as 'fair trade' have to <u>pay producers</u> a <u>fair price</u>.

3) <u>Buyers</u> also pay <u>extra</u> on top of the fair price to <u>help develop</u> the area where the goods come from, e.g. to <u>build schools</u> or <u>health centres</u>.

4) Only producers that <u>treat their employees well</u> can <u>take part</u> in the scheme. E.g. producers <u>aren't allowed to discriminate</u> based on sex or race, and employees must have a <u>safe working environment</u>. This <u>improves quality of life</u> for the employees.

5) However, producers in a fair trade scheme often <u>produce a lot</u> because of the good prices — this can cause them to produce too much. An <u>excess</u> will make <u>world prices fall</u> and cause producers who <u>aren't</u> in a fair trade scheme to <u>lose out</u>.

A 'fair price' is a price that's high enough for the producer to make a profit.

Trading groups

1) These are <u>groups of countries</u> that make <u>agreements</u> to <u>reduce barriers</u> to <u>trade</u> (e.g. to reduce import taxes) — this <u>increases</u> trade <u>between members</u> of the group.

2) When a poor country <u>joins</u> a trading group, the amount of <u>money</u> the country gets from trading <u>increases</u> — <u>more money</u> means that <u>more development</u> can take place.

3) However, it's <u>not easy</u> for poorer <u>countries that aren't part</u> of trading groups to <u>export goods</u> to <u>countries that are part</u> of trading groups. This <u>reduces</u> the <u>export income</u> of non-trading group countries and <u>slows down</u> their <u>development</u>.

E.g. NAFTA is a trade group including the USA, Canada and Mexico.

The Debt of Poorer Countries can be Reduced

1) <u>Debt abolition</u> is when some or all of a country's debt is <u>cancelled</u>. This means they can use the money they make <u>to develop</u> rather than to pay back the debt. For example, <u>Zambia</u> (in southern Africa) had <u>$4 billion</u> of <u>debt cancelled</u> in <u>2005</u>. In 2006, the country had enough money to start a <u>free healthcare</u> scheme for <u>millions of people</u> living in <u>rural areas</u>, which <u>improved</u> their <u>quality of life</u>.

2) <u>Conservation swaps</u> (debt-for-nature swaps) are when part of a country's debt is <u>paid off</u> by someone else in <u>exchange</u> for <u>investment</u> in <u>conservation</u>. For example, in 2008 the <u>USA</u> reduced <u>Peru's debt</u> by <u>$25 million</u> in exchange for <u>conserving</u> its <u>rainforests</u>.

Fair trade — weeks of revision and stress for a piece of paper with a grade on it...

Don't be glum, you're <u>over halfway</u> through this gem of a topic now. There's just a <u>little bit more revision</u> to go until you're a fully-fledged development guru — something I know you've wanted to be since you were little...

Reducing Global Inequality

Another big way <u>poorer countries</u> are given a <u>helping hand</u> is through <u>international aid</u>...

International Aid is Given from One Country to Another

1) Aid is <u>given</u> by one country to another country in the form of <u>money</u> or <u>resources</u> (e.g. food, doctors).

2) The country that <u>gives</u> the aid is called the <u>donor</u> — the one that <u>gets</u> the aid is called the <u>recipient</u>.

3) There are <u>two</u> main <u>sources</u> of aid from donor countries — <u>governments</u> (paid for by <u>taxes</u>)
   and <u>Non-Governmental Organisations</u> (NGOs, paid for by <u>voluntary donations</u>).

4) There are <u>two</u> different ways <u>donor governments</u> can give aid to recipient countries:

   > • <u>Directly</u> to the recipient — this is called <u>bilateral aid</u>.
   >
   > • <u>Indirectly</u> through an <u>international organisation</u>
   >   that distributes the aid — this is called <u>multilateral aid</u>.

   International organisations include the UN and the World Bank.

5) Bilateral aid can be <u>tied</u> — this means it's given with the <u>condition</u> that the <u>recipient country</u> has
   to <u>buy</u> the <u>goods and services</u> it needs <u>from</u> the <u>donor country</u>. This helps the <u>economy</u> of the
   donor country. However, if the goods and services are <u>expensive</u> in the donor country,
   the aid <u>doesn't go as far</u> as it would if the goods and services were bought <u>elsewhere</u>.

6) Aid can be classed as either <u>short-term</u> or <u>long-term</u> depending on <u>what it's used for</u>:

Type of Aid	What it is	Advantage to recipient	Disadvantage to recipient
<u>Short-term</u>	Money or resources to help recipient countries <u>cope</u> with <u>emergencies</u>, e.g. earthquakes, wars.	The <u>impact</u> of the aid will be <u>immediate</u> — <u>more people</u> will <u>survive</u> the emergency.	The <u>development</u> of a country remains <u>unchanged</u>. It may become <u>reliant on aid</u>.
<u>Long-term</u>	Money or resources to help recipient countries become <u>more developed</u>, e.g. to improve healthcare.	Countries will be <u>less reliant</u> on foreign aid as they become more developed.	It can <u>take a while</u> before the aid <u>benefits</u> a country, e.g. hospitals take a long time to be built.

7) Aid has one big <u>disadvantage</u> for <u>donor</u> countries — it <u>costs</u> them <u>money</u> or <u>resources</u>. However, one
   <u>advantage</u> is that the recipient countries become their <u>political allies</u> (they back them up on issues).

8) Some recipient countries <u>don't use aid effectively</u> because they have <u>corrupt governments</u> —
   the government uses the money and resources to <u>fund</u> their <u>lifestyle</u> or to pay for <u>political events</u>.

Long-Term Aid is Spent on Development Projects

Here are a couple of examples of the <u>type</u> of <u>development projects</u> aid is spent on:

> • Constructing <u>schools</u> to <u>improve literacy</u> rates, and <u>hospitals</u> to <u>reduce mortality</u> rates.
> • Building <u>dams</u> and <u>wells</u> to <u>improve clean water supplies</u>.
> • Providing <u>farming knowledge</u> and <u>equipment</u> to <u>improve agriculture</u>.

International Aid Donors Encourage Sustainable Development

<u>Sustainable development</u> means developing in a way that <u>doesn't irreversibly damage</u> the <u>environment</u> or
<u>use up resources</u> faster than they can be replaced. International aid donors (e.g. governments and NGOs)
<u>encourage</u> sustainable development in a number of ways:

What they do:	How it's sustainable:
<u>Invest</u> in <u>renewable energy</u>, to <u>reduce</u> the use of <u>fossil fuels</u>.	<u>Reduces</u> the <u>environmental impact</u> of using <u>fossil fuels</u>.
<u>Educate people</u> about <u>their environmental impact</u>.	<u>Reduces</u> things like <u>air</u> and <u>water pollution</u>.
<u>Plant trees</u> in areas that have been affected by <u>deforestation</u>.	Makes sure there are <u>still trees</u> to use <u>in the future</u>.

Watching Live Aid on DVD doesn't count as revision...

Quite a bit to remember on this page — try <u>drawing</u> a simple <u>diagram</u> to show all the <u>different ways</u> that <u>aid</u> goes from
a <u>donor</u> country to a <u>recipient</u> country. If you're still a bit confused after that just give Bob Geldof a call...

Reducing Global Inequality — Case Study

There are loads of <u>development projects</u> going on around the world to <u>reduce inequality</u>. Much as I'd like to, I can't possibly tell you about all of them, so here's an example of one...

FARM-Africa helps the Development of Rural Africa

1) <u>FARM-Africa</u> is a <u>non-governmental organisation</u> (NGO) that provides aid to <u>eastern Africa</u>.
2) It's funded by <u>voluntary donations</u>.
3) It was founded in <u>1985</u> to <u>reduce rural poverty</u>.
4) FARM-Africa runs programmes in <u>five</u> African countries — <u>Ethiopia</u>, <u>Sudan</u>, <u>Kenya</u>, <u>Uganda</u> and <u>Tanzania</u>.
5) FARM-Africa has been operating in <u>Ethiopia</u> since <u>1988</u>. Here are <u>four</u> of the projects it runs there:

Project	Region	Problem	What's being done	Helping...
<u>Rural Women's Empowerment</u>	Various	There are very <u>few opportunities</u> for Ethiopian women to <u>make money</u>. This means they have a <u>low quality of life</u> and <u>struggle to afford</u> things like <u>healthcare</u>.	Women are given <u>training</u> and <u>livestock</u> to <u>start farming</u>. <u>Loan schemes</u> have been set up to help women launch <u>small businesses</u> like <u>bakeries</u> and <u>coffee shops</u>. Women have been given <u>legal training</u> to <u>advise</u> other women of their <u>rights</u>.	Around <u>15 160 people</u>.
<u>Prosopis Management</u>	Afar	Prosopis, a <u>plant introduced</u> by the government to stabilise soils, has become a <u>pest</u> — it <u>invades grazing land</u>, making <u>farming difficult</u>.	Farmers are shown how to <u>convert prosopis</u> into <u>animal feed</u>. The animal feed is then <u>sold</u>, generating a new <u>source of income</u>.	Around <u>4400 households</u>.
<u>Community Development Project</u>	Semu Robi	Frequent <u>droughts</u> make <u>farming very difficult</u>. This <u>reduces</u> the <u>farmers' income</u> and can lead to <u>malnutrition</u>. Semu Robi is a <u>remote region</u> so getting <u>veterinary care</u> for livestock is <u>difficult</u>.	People are given <u>loans</u> to <u>buy</u> <u>small water pumps</u> to <u>irrigate</u> their farmland. This <u>reduces</u> the <u>effects of drought</u>. People are <u>trained</u> in <u>basic veterinary care</u> so they can help keep <u>livestock healthy</u>.	Around <u>4100 people directly</u> and <u>60 000 indirectly</u> by <u>veterinary care</u>.
<u>Sustainable Forest Management</u>	Bale	Forests <u>are cut down</u> to make land for <u>growing crops</u> and <u>grazing livestock</u>. Trees are also cut down for <u>firewood</u>. This <u>reduces resources</u> for <u>future generations</u>.	Communities are taught how to <u>produce honey</u> and <u>grow wild coffee</u>. These are then <u>sold</u>, so people can <u>make money without cutting down trees</u>. Communities are also taught how to make <u>fuel-efficient stoves</u> that <u>use less wood</u>. This also <u>reduces deforestation</u>.	Around <u>7500 communities</u>.

You'd need a seriously meaty plough to farm the whole of Africa...

Yes, it's <u>another case study</u> for you. I think this one's pretty interesting though, and it's not too difficult to get your head around. Make sure you can describe the <u>problems</u> in Ethiopia as well as the <u>development projects</u> that tackle them.

Inequalities in the EU — Case Study

You need to be able to describe the <u>factors</u> that have influenced the <u>level of development</u> of <u>two</u> contrasting <u>countries</u> in the EU, and describe some ways the EU has tried to <u>reduce</u> these inequalities.

Bulgaria is Less Developed than The UK

Bulgaria <u>joined</u> the EU in <u>2007</u> — it's <u>less developed</u> than the <u>UK</u>. For example:

- In 2007 Bulgaria had a <u>GNI per head</u> of <u>$11 180</u> and the UK had a GNI per head of <u>$33 800</u>.
- <u>Life expectancy</u> in Bulgaria is <u>six years lower</u> (<u>73</u> compared to <u>79</u>).
- The <u>HDI</u> for Bulgaria is <u>0.824</u>, whereas it's <u>0.947</u> for the UK.

Here are a <u>few reasons why</u> <u>Bulgaria</u> is less developed than the <u>UK</u>:

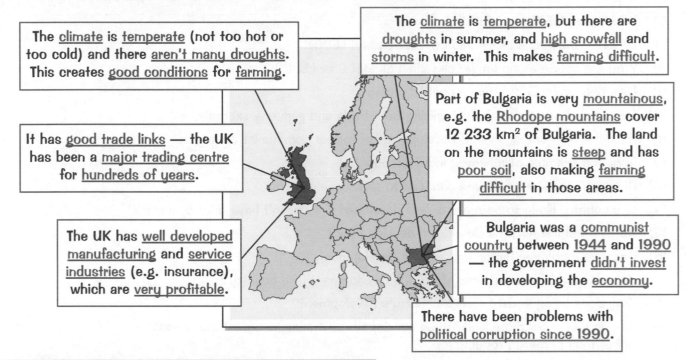

The <u>climate</u> is <u>temperate</u> (not too hot or too cold) and there <u>aren't many droughts</u>. This creates <u>good conditions</u> for <u>farming</u>.

The <u>climate</u> is <u>temperate</u>, but there are <u>droughts</u> in summer, and <u>high snowfall</u> and <u>storms</u> in winter. This makes <u>farming difficult</u>.

Part of Bulgaria is very <u>mountainous</u>, e.g. the <u>Rhodope mountains</u> cover 12 233 km² of Bulgaria. The land on the mountains is <u>steep</u> and has <u>poor soil</u>, also making <u>farming difficult</u> in those areas.

It has <u>good trade links</u> — the UK has been a <u>major trading centre</u> for <u>hundreds of years</u>.

The UK has <u>well developed manufacturing</u> and <u>service industries</u> (e.g. insurance), which are <u>very profitable</u>.

Bulgaria was a <u>communist country</u> between <u>1944</u> and <u>1990</u> — the government <u>didn't invest</u> in developing the <u>economy</u>.

There have been problems with <u>political corruption since 1990</u>.

The EU is Trying to Reduce Inequalities

Here are some of the ways that the EU is trying to <u>reduce inequality</u> in and between its member countries:

1) The <u>URBAN Community Initiative</u> — money is given to certain EU <u>cities</u> to <u>create jobs</u>, <u>reduce crime</u> and <u>increase</u> the area of <u>green space</u> (e.g. parks).

2) The <u>Common Agricultural Policy</u> (CAP) — farmers are <u>subsidised</u> (paid) to <u>grow certain products</u>. Also, when world food <u>prices</u> are <u>low</u>, the <u>EU buys produce</u> and guarantees farmers a reasonable income. The CAP also puts a <u>high import tax</u> on foreign produce so people in the EU are <u>more likely</u> to buy food <u>produced in the EU</u>. All these things <u>improve</u> the <u>quality of life</u> for farmers.

3) <u>Structural Funds</u> — these provide money for <u>research and development</u>, improving <u>employment opportunities</u>, <u>reducing discrimination</u> and improving <u>transport links</u>. The aim of the fund is to get <u>all members</u> of the EU to a <u>similar level</u> of <u>development</u> (reducing inequalities within Europe).

The EU is also trying to develop <u>Bulgaria</u> in some specific ways:

- <u>SAPARD</u> (Special Accession Programme for Agriculture and Rural Development) gives money to Bulgaria and two other countries to <u>invest</u> in <u>agriculture</u>.
- <u>Funds</u> earmarked for Bulgaria have been <u>partially frozen</u> until the <u>government</u> shows it's making <u>progress</u> in <u>fighting corruption</u>.

Bulgaria's not as developed as the UK but it produces champion sumo wrestlers...

...seriously, I'm not joking, who would've thought it. Well, you've <u>nearly made it to the end</u> of this topic, so you deserve a <u>mini treat</u> before you have a go at the revision summary — go and get a biscuit and a cup of tea.

Revision Summary for The Development Gap

Time to find out if you've developed a memory of development facts. Try these questions and if there are any that you don't know immediately have a look back at the page to refresh that big grey blob of yours. I know it's hard work, but it'll be worth it come the exam...

1) What is Gross National Income (GNI)?

2) Define birth rate.

3) What is the average age a person can expect to live to called?

4) Give three things that are used to calculate the HDI for a country.

5) A single measure of development could be used to figure out how developed a country is. Give one limitation of this.

6) What does MEDC stand for?

7) Describe the global distribution of MEDCs and LEDCs.

8) Why are countries no longer classified as MEDC or LEDC?

9) Give one example of a rich industrial country.

10) Describe what a Newly Industrialising Country is and give one example.

11) Describe the general level of development of former communist countries.

12) Give one example of a heavily indebted country.

13) Give one way a poor climate can lead to slow development.

14) Is a country likely to be more or less developed if it doesn't have a lot of water?

15) How do natural hazards slow down development?

16) Describe a political factor that affects development.

17) How do world trade patterns affect the development of a country?

18) Why does being in debt slow a country's development?

19) Why is a country more likely to be developed if it trades manufactured goods rather than primary products?

20) a) Name a natural disaster that set back a country's development.

   b) Describe the effects that the disaster had on development.

21) Give two ways that people in poor countries are trying to improve their own quality of life.

22) What is fair trade?

23) How do trading groups help to reduce inequalities?

24) Give two ways the debt of a poor country can be reduced.

25) What is multilateral aid?

26) Give one advantage of short-term aid.

27) Give one advantage and one disadvantage of international aid for donor countries.

28) Give one way that international aid donors encourage sustainable development.

29) a) Name a development project.

   b) Describe how the development project is attempting to reduce global inequality.

30) a) Name two countries in the EU that have contrasting levels of development.

   b) State three reasons why they have different levels of development.

   c) Give two ways in which the EU is trying to reduce the differences.

Globalisation Basics

Globalisation is a long word and a big topic. Better get started then...

Globalisation is the Process of Economies Becoming More Integrated

1) Globalisation is the process of all the world's economies becoming integrated
   — it's the whole world coming together like a single community.

2) It happens because of international trade, international investment
   and improvements in communications.

3) Countries have become interdependent as a result of globalisation
   — they rely on each other for resources or services.

Globalisation is also about cultures and political policies becoming more integrated.

Improvements in Communications have Increased Globalisation

Improvements in ICT (Information and Communication Technology) and
transport have increased globalisation by increasing trade and investment:

ICT

1) Improvements in ICT include e-mail, the internet, mobile phones and
   phone lines that can carry more information and faster.

2) This has made it quicker and easier for businesses all over the world
   to communicate with each other. For example, a company can have
   its headquarters in one country and easily communicate with branches
   in other countries. No time is lost so it's really efficient.

Transport

1) Improvements in transport include more airports, high-speed trains and larger ships.

2) This has made it quicker and easier for people all over the world to communicate
   with each other face to face.

3) It's also made it easier for companies to get supplies from all over the world,
   and to distribute their product all over the world. They don't have to be located
   near to their suppliers or their product market anymore.

These improvements have allowed the development of call centres abroad and localised industrial regions:

Call centres abroad

1) Call centres are used by some companies
   to handle telephone enquiries about
   their business.

2) Improvements in ICT mean that it's
   just as easy for people to phone a
   faraway country as it is to phone
   people in their own country.

3) So a lot of call centres are now based
   abroad because the labour is cheaper,
   which reduces running costs.

EXAMPLE: In 2004 Aviva (an insurance
company) moved 950 call centre jobs from
the UK to India and Sri Lanka, as it costs
less there (e.g. it costs 40% less in India).

Localised industrial regions

Improvements in ICT and transport have allowed
some industries to develop around a specific
region that's useful to them, but still have global
connections to get all the other things they need.

EXAMPLE: A lot of motorsport companies have
offices in Oxfordshire and Northamptonshire,
e.g. the Renault Formula 1 team have their
headquarters there. They're close to the
Silverstone race circuit (so they can test their
cars) and the area has lots of skilled workers.
People like drivers and engineers can easily fly
into the area. The manufacturers use the internet
to easily send and receive information and data
about their cars to people around the world.

Globalisation is going large — with extra barbecue sauce...

Globalisation is a bit of a weird concept so don't panic if you don't get it straight away. Learn how improvements in ICT
and transport have increased globalisation, and don't forget an example for call centres and localised industrial regions.

Trans-National Corporations (TNCs)

You can't get too far into the globalisation topic before stumbling across Trans-National Corporations.

TNCs also Increase Globalisation

1) TNCs are companies that produce products, sell products or are located in more than one country. For example, Sony is a TNC — it manufactures electronic products in China and Japan.

2) They increase globalisation by linking together countries through the production and sale of goods.

3) TNC factories are usually located in poorer countries because labour is cheaper, which means they make more profit (see the next page for more reasons why they're located in poorer countries).

4) TNC offices and headquarters are usually located in richer countries because there are more people with administrative skills (because education is better).

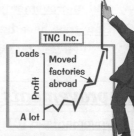

5) TNCs have advantages and disadvantages:

Advantages	Disadvantages
TNCs create jobs in all the countries they're located in.	Employees in poorer countries may be paid lower wages than employees in richer countries.
Employees in poorer countries get a more reliable income compared to jobs like farming.	Employees in poorer countries may have to work long hours in poor conditions.
TNCs spend money to improve the local infrastructure, e.g. airports and roads.	Most TNCs come from richer countries so the profits go back there — they aren't reinvested in the poorer countries they operate in.
New technology (e.g. computers) and skills are brought to poorer countries.	The jobs created in poorer countries aren't secure — the TNC could relocate the jobs to another country at any time.

Case Study — Wal-Mart® is a Retail TNC with Headquarters in the USA

1) Wal-Mart began in 1962 when Sam Walton opened the first store in Arkansas, USA.

2) More stores opened across Arkansas, then across the USA, and more recently across the world, e.g. in Mexico, Japan, Brazil, Canada and the UK (where it's called ASDA).

3) Wal-Mart sells a variety of products, e.g. food, clothes and electrical goods.

4) Wal-Mart is the biggest retailer in the world — it owns over 8000 stores and employs over 2 million people.

5) Wal-Mart has a number of advantages and disadvantages, for example:

Advantages of Wal-Mart	Disadvantages of Wal-Mart
Every new store that's built creates jobs, e.g. a new store in Vineland (USA) opened in 2009, creating 700 jobs.	Factory workers in the USA earn around $6 an hour, but factory workers in China earn less than $1 an hour (though this is quite a lot in China).
Wal-Mart donates hundreds of millions of dollars to improve education, healthcare and the environment in the countries that it's based in.	Some companies that supply Wal-Mart have long working hours, e.g. Beximco in Bangladesh supplies clothing. Bangladesh has a maximum 60 hour working week, but some people claim employees at Beximco regularly work 80 hours a week.

TNCs are everywhere — and I mean everywhere...

Geography types are always making up long names for things (trans-national corporations), then squishing them down to just a few letters (TNC). I don't know what they get out of it but it seems to keep them happy.

Change in Manufacturing Location

TNCs can put their factories anywhere in the world, but some countries are more attractive than others. This has meant some countries are now stuffed full of factories and others are waving bye bye to them.

The Manufacturing Industry is Growing in Some Countries...

1) Some countries that have traditionally relied on agriculture have seen a massive growth in their manufacturing industries recently — this process is called industrialisation.

2) These countries are called NICs (Newly Industrialising Countries). They include places like India, China and Brazil.

An increase in manufacturing creates jobs.

3) TNCs have increased manufacturing in NICs by basing factories there — here are five reasons why they do this:

1 Cheap labour

The minimum wage is the lowest amount a company is allowed to pay someone. It's set by governments. Some NICs don't have a minimum wage. In the ones that do it's much lower than in richer countries, e.g. the UK. This reduces the cost of manufacturing goods because factory workers are paid less.

2 Long working hours

The rules about working hours aren't as strict in NICs as in places like the EU. This means employees work longer hours so more product can be made in a day.

3 Laxer health and safety regulations

There are fewer health and safety regulations in NICs and they're often not enforced. This lowers the cost of manufacturing goods because less money is spent on increasing the safety of factories.

4 Prohibition of strikes

Some NICs don't allow employees to strike, e.g. to protest against low pay. This means money isn't lost due to employees stopping work.

5 Tax incentives and tax free zones

Some NICs offer TNCs a tax reduction if they move their manufacturing to the country. Some NICs have tax free zones — the TNCs don't have to pay taxes if they move their manufacturing to a specific area of the country. Both of these increase the profits of the TNC because they pay lower taxes.

...and Declining in Other Countries

1) Most rich countries have a history of manufacturing goods, e.g. cars have been manufactured in the UK for many years. In recent years manufacturing in some rich countries has decreased.

2) This process is called deindustrialisation.

3) Deindustrialisation can happen for a number of reasons, for example:

- Manufacturers move factories abroad because they can produce goods more cheaply there.
- Manufacturers close down because they can't compete with the price of goods manufactured abroad.

4) When deindustrialisation happens a lot of manual workers (e.g. factory and dock workers) lose their jobs. Also, as factories close some buildings become derelict. But, there's often an increase in service industries like banking and insurance. These industries pay people higher wages than manufacturing so deindustrialisation isn't all bad.

Long hours and no striking allowed — sounds like a typical week revising...

The basic formula is 'more factories here = less factories there'. Get that fact fixed in your grey matter, then learn the reasons why it's the height of fashion to put your factory in a NIC. Don't forget about deindustrialisation either.

Change in Manufacturing Location — Case Study

You're not getting away that easily — here's another <u>case study</u> for you. You've got to know about <u>China's development</u> into an <u>economic giant</u> and the <u>reasons</u> why <u>manufacturing</u> is <u>moving to the country</u>.

China is one of the World's Fastest Growing Industrial Economies

1) In <u>30 years</u> China has <u>gone from</u> being a <u>mainly agricultural economy</u> to a <u>strong manufacturing economy</u>. It's now the <u>third largest economy</u> in the world after the US and Japan.

2) The <u>percentage of China's GDP</u> that came from <u>agriculture fell</u> between <u>1978</u> and <u>2004</u>, from about <u>30%</u> to <u>less than 15%</u>.

> GDP is the total value of goods and services a country produces in a year.

3) During the <u>same time</u> the number of <u>products manufactured</u> in China has <u>increased rapidly</u>, e.g. about <u>4000 colour TVs</u> were made in China in <u>1978</u> compared to nearly <u>75 million</u> in <u>2004</u>.

4) China manufactures <u>loads of different products</u> like <u>clothes</u>, <u>computers</u> and <u>toys</u>.

5) Lots of <u>TNCs</u> have <u>factories in China</u>, for example <u>NIKE</u>, <u>Hewlett-Packard</u> and <u>Disney</u>.

There are Lots of Reasons for the Growth in Manufacturing

1 Cheap labour

There's <u>no single minimum wage</u> in China — it's <u>different all over the country</u>. For example, in <u>Shenzhen</u> the minimum wage is about <u>£90 per month</u> and in <u>Beijing</u> it's about <u>£70 per month</u>. This makes <u>labour</u> in China <u>much cheaper</u> than other countries, e.g. in the <u>UK</u> the <u>minimum wage</u> is about <u>£990 per month</u>.

2 Long working hours

<u>Chinese law</u> says that people are only allowed to work <u>40 hours per week</u>, with a maximum of <u>36 hours</u> of <u>overtime per month</u>. This <u>isn't always enforced</u> though — for example, the manufacturing company <u>foxconn®</u> said that some of its Chinese factory workers have done about <u>80 hours of overtime per month</u> to <u>maximise</u> the <u>production</u> of goods.

3 Laxer health and safety regulations

The <u>health and safety laws</u> in China are <u>similar to other countries</u> but they <u>aren't heavily enforced</u>, e.g. over the <u>past decade</u>, <u>hundreds</u> of factory workers have been <u>treated for mercury poisoning</u> despite <u>strict laws</u> on <u>working with toxic materials</u>.

4 Prohibition of strikes

Chinese workers <u>can</u> go on <u>strike</u> but the <u>All-China Federation of Trade Unions</u> (ACFTU) is <u>required by law</u> to get people <u>back to work</u> as <u>quickly as possible</u> so <u>productivity</u> is <u>maximised</u>. It's <u>illegal</u> for people to <u>join any union other than the ACFTU</u>.

5 Tax incentives and tax free zones

China has many <u>Special Economic Zones</u> (SEZs) that offer <u>tax incentives</u> to <u>foreign businesses</u>. <u>Foreign manufacturers</u> usually pay <u>no tax</u> for the <u>first two years</u> in the zone, <u>7.5%</u> for the <u>next three years</u> and then <u>15% from then on</u> (which is still <u>half</u> of the <u>usual 30%</u> tax <u>elsewhere</u> in China). <u>Shenzhen</u> is one of the <u>most successful SEZs</u>. There's been around <u>$30 billion</u> of <u>investment by TNCs</u>. Factories in Shenzhen <u>make products</u> for companies like <u>Wal-Mart®</u>, <u>Dell™</u> and <u>IBM®</u>.

Fe fi fo fum — China gets the world's manufacturing done...

I bet you're within about a metre of something that was made in China — unless you're doing your revision on the surface of the Sun. Make sure you can give all the <u>reasons</u> why <u>manufacturing in China</u> has <u>gone through the roof</u>.

Globalisation and Energy Demand

All this globalisation has made us hungry. No, not for courgettes, but for energy...

The Global Demand for Energy is Increasing

Globalisation has increased the wealth of some poorer countries so people are buying more things. A lot of these things use energy, e.g. cars, fridges and televisions. This increases the global demand for energy. There are two other reasons why the global demand for energy is increasing:

1) Technological advances have created loads of new devices that all need energy, e.g. computers, mobile phones and MP3 players. These are becoming more popular so more energy is needed.

2) In 2000 the world population was just over 6 billion and it's projected to increase to just over 9 billion in 2050 — more people means more energy is needed.

Producing More Energy has Lots of Impacts

Most of the energy produced in the world comes from burning fossil fuels (i.e. oil, gas and coal). Nuclear power, wood and renewable sources (e.g. solar power) are also used to produce some energy. Increasing energy production to meet demand has social, economic and environmental impacts:

Social impacts

1) More power plants will have to be built to increase energy production. Power plants are extremely large — people may have to move out of an area so a power plant can be built.

2) The waste from nuclear power plants is radioactive. If it leaks out from where it's stored it can cause death and illness, and can contaminate large areas of land. If more nuclear power plants are built to increase energy production, there's a higher risk of radioactive waste leaking out.

3) Increasing energy production will create jobs — people will be needed to build more power stations, run them and maintain them.

Environmental impacts

1) Burning fossil fuels releases carbon dioxide (CO_2). This adds to global warming. Global warming will cause the sea level to rise, cause more severe weather and force species to move (to find better conditions) or make them extinct (if they can't move and it gets too hot). Using more fossil fuels will increase global warming.

Economic impact

Countries with lots of energy resources, e.g. lots of coal, will become richer as energy demand increases — countries with few resources will need to buy energy from them.

2) Burning fossil fuels also releases other gases that dissolve in water in the atmosphere and cause acid rain. Acid rain can kill animals and plants. Using more fossil fuels will increase acid rainfall.

3) Gathering wood for fuel can cause deforestation (removing trees from forests). Removing trees destroys habitats for animals and other plants. Using more wood for fuel will increase deforestation.

4) Mining for coal causes air and water pollution. It also removes large areas of land, which destroys habitats. More coal mining will cause more pollution and destroy more habitats.

5) Transporting oil is a risky business — oil pipes and tankers can leak, spilling oil. Oil spills can kill birds and fish. Using more oil means more needs to be transported, increasing the risk of spills.

Maybe I'll turn my computer off standby then...

Crikey, producing energy has a lot of impacts and the impacts are bigger the more energy you produce. Make sure you learn them so you can trot them out in nice neat handwriting for the examiner (if it's illegible you won't get any marks).

Globalisation and Food Supply

I'm quite partial to <u>strawberries</u> with my <u>Christmas cake</u>, and it seems I'm not the only one...

Food Production has become Globalised

1) Before the 1960s people mainly ate a <u>small range</u> of <u>seasonal food</u> that had been <u>grown</u> in their <u>own country</u> (often in their <u>local area</u>).

2) People now <u>demand</u> to have a <u>range</u> of foods <u>all year round</u>, regardless of growing seasons. This has led to <u>globalisation</u> of the food industry — food is <u>produced</u> in <u>foreign countries</u> and <u>imported</u>.

3) The <u>increase</u> in the <u>world's population</u> also means <u>more food is needed</u> — the <u>demand has increased</u>.

4) Countries are trying to <u>increase food production</u> to meet this demand, but <u>some</u> <u>can't produce enough</u> to feed their population so food has to be <u>imported</u> too.

Producing More Food and Importing Food has Lots of Impacts

Environmental

1) <u>Transporting</u> food <u>produces CO$_2$</u>. The <u>distance food is transported to the market</u> is called <u>food miles</u>. The <u>higher the food miles</u>, the <u>more CO$_2$</u> is produced. CO$_2$ adds to <u>global warming</u>.

2) The <u>amount of CO$_2$ produced</u> during <u>growing and transporting</u> a food is called its <u>carbon footprint</u>. A <u>larger</u> carbon footprint means <u>more CO$_2$</u> and <u>more global warming</u>.

3) <u>Imported foods</u> have to be <u>transported a long way</u> so have <u>high food miles</u> and a <u>large carbon footprint</u>. A <u>benefit</u> of importing food is that a <u>wide range of food</u> is available <u>all year round</u>. Another benefit is it <u>helps meet increasing demand</u> in countries that can't produce a lot.

4) <u>More food</u> could be <u>produced locally</u> by <u>energy intensive farming</u> — <u>pesticides</u>, <u>fertilisers</u> and <u>machinery</u> are used to <u>produce large quantities</u> of food. Although <u>food miles</u> are <u>low</u>, <u>loads of energy</u> is needed to <u>make chemicals</u> and <u>run the machinery</u>. <u>Energy production</u> <u>creates lots of CO$_2$</u> so local energy intensive farming can have a <u>large carbon footprint</u>.

5) To <u>produce more food</u> some farmers use <u>marginal land</u> (land that's <u>not really suitable</u> for farming), e.g. steep hillsides or the edges of deserts. The <u>soil</u> in marginal land is <u>thin</u> and it's <u>quickly eroded</u> by farming, <u>degrading the environment</u>.

Political

<u>Lots of water</u> is needed to produce <u>lots of food</u>. Farmers in countries with <u>low rainfall</u> need to <u>irrigate</u> their land with water from <u>rivers and lakes</u>. As the <u>demand</u> for <u>water increases</u> (due to the increased demand for food) there may be <u>hostilities between countries</u> that <u>use the same water source</u> for irrigation. For example, there's <u>tension</u> between <u>Egypt</u>, <u>Sudan</u> and <u>Ethiopia</u> because they all take water from the <u>River Nile</u>.

Social

Some farmers are <u>switching</u> from <u>subsistence farming</u> (where food is produced for <u>their family</u>) to <u>commercial farming</u> (where food is produced to <u>sell</u>). This is because they can <u>make more money</u> due to the high demand for food. This <u>reduces</u> the <u>amount</u> of <u>food produced</u> for <u>local people</u> so they have to <u>import food</u> (which is <u>more expensive</u>). If <u>food prices go down</u>, then farmers might <u>not earn enough money</u> to <u>buy food</u> for themselves.

Crops sold to make money are called cash crops.

Economic

1) <u>Using chemicals</u> (e.g. <u>fertilisers</u>, <u>pesticides</u> and <u>insecticides</u>) helps to <u>produce lots of food</u>. These chemicals can be <u>very expensive</u> — farmers may have to <u>borrow money</u> to buy the chemicals and this <u>gets farmers into debt</u>.

2) Farmers can generate a <u>steady income</u> by <u>producing food</u> for <u>export</u> to other countries.

Eat globally — don't stop till you're a perfect sphere...

Farming used to be a totally <u>local business</u> but nowadays our <u>food comes flying in</u> from all over the world (or trundling in on a lorry, train or container ship). There are <u>more mouths to feed</u> too. Make sure you know all the <u>knock-on effects</u>.

Reducing the Impacts of Globalisation

One impact of globalisation is that more people are gluttons for energy. <u>Producing more energy</u> using fossil fuels has <u>plenty of impacts</u>, but worry not, things can be done to make <u>producing energy greener</u>.

Using Renewable Energy Is a Sustainable Way to Meet Energy Demands

1) Energy production needs to be <u>sustainable</u> — it needs to allow people alive <u>today</u> to get what they <u>need</u> (energy), but <u>without stopping people</u> in the <u>future</u> getting what they <u>need</u>. This basically means <u>not damaging the environment</u> or <u>using up resources</u> faster than they can be replaced.

2) Producing energy using <u>fossil fuels</u> (i.e. coal, oil and gas) <u>isn't sustainable</u>.

3) This is because fossil fuels are <u>non-renewable</u> — this means they'll <u>eventually run out</u> so there <u>won't be any</u> for <u>future generations</u>.

Wander back to p. 117 for more on the impacts of producing energy.

4) Using fossil fuels also <u>damages the environment</u>, e.g. <u>burning them</u> <u>produces CO_2</u>, which <u>causes global warming</u>.

5) Energy produced from <u>renewable sources</u> is <u>sustainable</u> because it <u>doesn't cause long term environmental damage</u> and the <u>resource won't run out</u>. Here are some renewable energy sources:

- <u>Wind</u> — the <u>wind turns blades</u> on a <u>wind turbine</u> to <u>generate electrical energy</u>.

- <u>Biomass</u> — biomass is <u>material</u> that comes from organisms that <u>are alive</u> (e.g. <u>animal waste</u>) or <u>were recently alive</u> (e.g. <u>plants</u>). It can be <u>burnt</u> to <u>release energy</u>. It can also be <u>processed</u> to produce <u>biofuels</u>, which are then <u>burnt</u> to <u>release energy</u>.

- <u>Solar power</u> — <u>energy from the sun</u> can be used to <u>heat water</u>, <u>cook food</u> and <u>generate electrical energy</u>.

- <u>Hydroelectric power</u> — <u>water</u> is <u>trapped behind a dam</u> and <u>forced through tunnels</u>. The water <u>turns turbines</u> in the tunnels to <u>generate electrical energy</u>.

6) Producing energy from renewable sources <u>contributes</u> to <u>sustainable development</u> — it allows areas to develop (i.e. use more energy to improve the lives of the people there) in a sustainable way.

Case Study — Spain is Using Wind Energy to Meet Demand

1) Spain's <u>energy consumption</u> has <u>increased 66% since 1990</u>.

2) <u>Some</u> of the extra energy needed is being produced using <u>wind turbines</u> — the amount of energy produced from wind has <u>increased 16-fold since 1995</u>.

3) Spain is <u>ideal</u> for wind farms because it has <u>large, windy areas</u> where <u>not many people live</u>. This means wind farms can be built <u>without annoying too many people</u>.

4) Spain has <u>over 400 wind farms</u> and a total of <u>over 12 000 turbines</u>.

5) In <u>2008</u>, <u>11.5%</u> of Spain's energy was supplied by <u>wind energy</u>.

Wind farms are groups of wind turbines.

6) The wind farms have had positive and negative <u>impacts</u>:

Positive impacts	Negative impacts
1) In <u>2008</u>, using wind energy <u>reduced</u> Spain's CO_2 emissions by over <u>20 million tonnes</u>.	1) <u>Some conservationists</u> say the wind farms are a <u>danger</u> to <u>migrating birds</u>.
2) In <u>2008</u>, using wind energy <u>saved</u> Spain from <u>importing</u> about €1.2 billion of <u>gas</u> and <u>oil</u>.	2) <u>Some people</u> think wind farms are <u>ugly</u> — turbines can be <u>seen from miles away</u>.
3) Spain's <u>wind energy industry</u> has created around <u>40 000 jobs</u>.	3) <u>Some people</u> think the wind farms are <u>too noisy</u>.

El stinky — a Spanish wind farm...

Sorry, I've sunk to lowly fart jokes after that page. It's not the most thrilling stuff in the world, but unfortunately you need to know how <u>renewable energy</u> is <u>sustainable</u> and <u>one case study</u> of its use. Happy revising.

Reducing the Impacts of Globalisation

Last page, last page, last page — you've only got to learn a few more ways to <u>reduce</u> the <u>impacts</u> of <u>globalisation</u> then you're done and dusted with this topic...

The Kyoto Protocol helps to Reduce Carbon Dioxide Emissions

1) Globalisation has <u>increased</u> the <u>demand for energy</u> (see p. 117) — <u>more fossil fuels</u> are being <u>used</u> to meet the demand, producing <u>loads of CO_2</u> and adding to <u>global warming</u>.

2) The <u>international community</u> is <u>working together</u> to <u>reduce</u> the amount of <u>CO_2</u> they produce because the <u>problem</u> of global warming <u>affects everyone</u>.

3) The <u>Kyoto Protocol</u> is an <u>international agreement</u> that has been signed by most countries in the world to <u>cut emissions of CO_2</u> and other gases by <u>2012</u>. Each country is set an <u>emissions target</u>, e.g. the <u>UK</u> has agreed to reduce emissions by <u>12.5%</u> by 2012.

4) Another part of the protocol is the <u>carbon credits trading scheme</u>:

> - <u>Countries</u> that come under their emissions target get <u>carbon credits</u> which they can <u>sell</u> to countries that <u>aren't meeting</u> their emissions target. This means there's a <u>reward</u> for having <u>low emissions</u>.
> - <u>Countries</u> can also <u>earn</u> carbon credits by helping <u>poorer countries</u> to <u>reduce</u> their emissions. This means poorer countries will be able to reduce their emissions <u>more quickly</u>.

International agreements are also called international directives.

Other International Agreements help to Reduce Pollution

1) Globalisation has <u>increased</u> the <u>emission of gases</u> that <u>cause pollution</u> like <u>acid rain</u> (see p. 117).

2) There are <u>international agreements</u> that help to <u>reduce pollution</u>, e.g. the <u>Gothenburg Protocol</u>.

3) The Gothenburg Protocol sets <u>emissions targets</u> for <u>European countries</u> and the <u>US</u>. The protocol aims to <u>cut harmful gas emissions</u> by <u>2010</u> to <u>reduce acid rain</u> and <u>other pollution</u>.

Recycling Reduces Waste Created by Globalisation

1) Globalisation means people have access to <u>more products</u> at <u>low prices</u>, so they can afford to be <u>more wasteful</u>, e.g. people <u>throw away</u> damaged clothes <u>instead of repairing</u> them.

2) Things that are <u>thrown away</u> get taken to <u>landfill sites</u> — the <u>amount of waste</u> going to landfill has <u>increased</u> as globalisation has increased.

3) One way to <u>reduce</u> this <u>impact</u> on a <u>local scale</u> is to <u>recycle waste</u> to make <u>new products</u>, e.g. recycling <u>old drinks cans</u> to make <u>new ones</u>.

Buying Local can Reduce the Impacts of a Globalised Food Supply

1) In recent years <u>celebrity chefs</u>, <u>food writers</u> and <u>campaigners</u> have encouraged people to <u>eat more locally-produced food</u>.

2) Buying local food helps to <u>reduce food miles</u> (see p. 118) because it <u>hasn't</u> been <u>transported a long way</u>. It also helps to <u>support local farmers</u> and <u>businesses</u>.

3) However, if people <u>only buy locally</u> it can put <u>people in poorer countries</u> who <u>export food</u> <u>out of a job</u>.

You're not local, are you...

And the Oscar® for the best directive goes to — the Gothenburg Protocol...

You know I said that once you've done this page, that's it for <u>globalisation</u>... well, it might have been a tiny lie. When you've learnt this page you can scoot on to the <u>revision summary</u>. After that, I promise you the topic is done...

Revision Summary for Globalisation

Globalisation... not the most cheery or thrilling of topics (for drama and excitement give me volcanoes any day of the week) — but still kind of important for life, the future of the planet and all that jazz. Before you put it all behind you and move on to something more uplifting, check you've got the hang of the main issues with this useful bunch of bananas. I mean questions. Sorry.

1) What is globalisation?

2) How have improvements in ICT increased globalisation?

3) How have improvements in transport increased globalisation?

4) Why are call centres often based abroad?

5) What is a TNC?

6) How do TNCs increase globalisation?

7) a) Name a TNC.

    b) Name two countries it is located in.

    c) Give one advantage of the TNC.

    d) Give one disadvantage of the TNC.

8) What is industrialisation?

9) What are NICs? Name one NIC.

10) Give two reasons why TNCs move to NICs.

11) Why does deindustrialisation happen?

12) a) Describe how China's economy has changed over the last 30 years.

    b) Explain how tax incentives have caused manufacturers to move to China.

    c) Give two other reasons why manufacturers move to China.

13) Explain how globalisation has increased the demand for energy.

14) Give two other things that have caused an increase in the demand for energy.

15) Give two environmental impacts of producing more energy.

16) Give an economic impact of producing more energy.

17) Give two reasons why food is imported into a country.

18) Describe a political impact of producing more food.

19) Describe a social impact of switching from subsistence farming to commercial farming.

20) What is sustainable energy production?

21) Why are fossil fuels not a sustainable energy source?

22) Give two examples of renewable energy sources.

23) a) Name a country that uses renewable energy.

    b) How much of the energy used in that country comes from renewable sources?

    c) Give one positive impact on that country of using renewable energy.

24) What is the Kyoto Protocol?

25) How does a country get carbon credits?

26) Name an international agreement that aims to reduce pollution.

27) Globalisation has increased the amount of waste going to landfill. How can this impact be reduced?

28) Give one reason why buying locally produced food reduces the impact of a globalised food supply.

Unit 2B — Tourism

Growth in Tourism

Examiners have a <u>twisted sense of humour</u> at times. As if it's not bad enough that you have to <u>stay inside</u> and <u>do revision all day</u>, just to rub it in they make you learn about <u>people going on holiday</u>.

There's been a Global Increase in Tourism Over the Last 60 Years

Tourism's a <u>growing industry</u> — people are having <u>more holidays</u> and <u>longer holidays</u>.
Here are a few of the <u>reasons why</u>:

1) People have <u>more disposable income</u> (spare cash) than they used to, so <u>can afford</u> to go on <u>more holidays</u>.
2) Companies give <u>more paid holidays</u> than they used to.
    This means people have <u>more free time</u>, so <u>go on holiday more</u>.
3) <u>Travel</u> has become <u>cheaper</u> (particularly <u>air travel</u>) so <u>more people</u> can <u>afford to go on holiday</u>.
4) <u>Holiday providers</u>, e.g. tour companies and hotels, now use the <u>internet</u> to <u>sell</u> their products to people <u>directly</u>, which makes them <u>cheaper</u>. Again, this means <u>more people can afford</u> to <u>go away</u>.

<u>Some areas</u> are also becoming <u>more popular</u> than they used to be because:

1) <u>Improvements in transport</u> (e.g. more airports) have made it <u>quicker</u> and <u>easier</u> to <u>get to places</u> — no more week-long boat trips to Australia for a start.
2) Countries in more <u>unusual tourist destinations</u> like the Middle East and Africa have got <u>better</u> at <u>marketing themselves</u> as tourist attractions. This means people are <u>more aware of them</u>.
3) Many countries have <u>invested</u> in <u>infrastructure for tourism</u> (e.g. better hotels) to make them <u>more attractive to visitors</u>.

Cities, Mountains and Coasts are all Popular Tourist Areas

People are attracted to <u>cities</u> by the <u>culture</u> (e.g. museums, art galleries), <u>entertainment</u> (bars, restaurants, theatres) and <u>shopping</u>. Popular destinations include <u>London</u>, <u>New York</u>, <u>Paris</u> and <u>Rome</u>.

People are attracted to <u>mountain areas</u> by the beautiful <u>scenery</u> and activities like <u>walking</u>, <u>climbing</u>, <u>skiing</u> and <u>snow boarding</u>. Popular destinations include the <u>Alps</u>, the <u>Dolomites</u> and the <u>Rockies</u>.

People are attracted to coastal areas by the <u>beaches</u> and activities like <u>swimming</u>, <u>snorkelling</u>, <u>fishing</u> and <u>water skiing</u>. Popular destinations include <u>Spain</u>, the <u>Caribbean</u> and <u>Thailand</u>.

Tourism is Important to the Economies of Many Countries

1) Tourism <u>creates jobs</u> for local people (e.g. in restaurants and hotels), which helps the <u>economy to grow</u>.
2) It also <u>increases the income</u> of <u>other businesses</u> that <u>supply the tourism industry</u>, e.g. farms that supply food to hotels. This also helps the <u>economy to grow</u>.
3) This means tourism is important to the economy of countries in both <u>rich</u> and <u>poor parts</u> of the world, e.g. tourism in France generated <u>35 billion Euros</u> in 2006 and created <u>two million jobs</u>.
4) Poorer countries tend to be <u>more dependent</u> on the income from tourism than richer ones, e.g. tourism contributes <u>3%</u> of the <u>UK's GNP</u>, compared to <u>15%</u> of <u>Kenya's</u>.

Tourism growth — you should probably have that removed...

Make sure you can recite the <u>reasons</u> why tourism is <u>on the increase</u> backwards, whilst standing on your head. And don't forget — if an area is <u>pretty</u> or has <u>tons of activities</u> then it'll be a <u>hit with tourists</u> (which is great for the <u>economy</u>).

UK Tourism

You might not realise it on yet <u>another rainy day</u> here, but the <u>UK</u> is actually a <u>top tourist destination</u>.

Tourism makes a Big Contribution to the UK Economy

1) There were <u>32 million overseas visitors</u> to Britain in <u>2008</u>.

2) The UK is popular with tourists because of its <u>countryside</u>, <u>historic landmarks</u> (e.g. Big Ben and Stonehenge), famous <u>churches and cathedrals</u> (e.g. Saint Paul's cathedral), and its <u>castles and palaces</u> (e.g. Edinburgh Castle and Buckingham Palace).

Really popular areas are called honeypot sites.

Welcome to the UK, old bean.

Aw, you Brits are so cute!

snap

3) <u>London</u> is particularly popular for its museums, theatres and shopping. It's the destination for <u>half of all visitors</u> to the UK.

4) In <u>2007</u>, tourism contributed <u>£114 billion</u> to the <u>economy</u> and <u>employed 1.4 million people</u>.

Many Factors Affect the Number of Tourists Visiting the UK

1) <u>Weather</u> — <u>bad weather</u> can <u>discourage tourists</u> from visiting the UK, e.g. a really <u>wet summer</u> in <u>2007</u> was blamed for a <u>drop</u> in the number of overseas visitors.

2) <u>World economy</u> — in times of <u>recession</u> people tend to <u>cut back</u> on <u>luxuries</u> like holidays, so <u>fewer overseas visitors</u> come to the UK. It's not all bad though, as <u>more UK citizens</u> choose to <u>holiday in the UK</u>.

3) <u>Exchange rate</u> — the <u>value</u> of the <u>pound</u> compared with other currencies affects the number of tourists. If it's <u>low</u>, the UK is <u>cheaper to visit</u> so more overseas visitors come.

4) <u>Terrorism and conflict</u> — wars and terrorist threats mean people are <u>less willing to visit</u> <u>affected areas</u>. <u>Tourism fell sharply</u> after the <u>London bombings</u> on <u>7th July 2005</u>.

5) <u>Major events</u> — big events can <u>attract huge numbers of people</u>. E.g. <u>Liverpool</u> was <u>European Capital of Culture</u> in <u>2008</u> and as a result <u>3.5 million</u> people visited that <u>hadn't been before</u>. The <u>2012 Olympics</u> are also expected to massively boost tourism.

The Tourist Area Life Cycle Model Shows How Visitor Numbers Change

The <u>number of visitors</u> to an area over time tends to go through these <u>typical stages</u>:

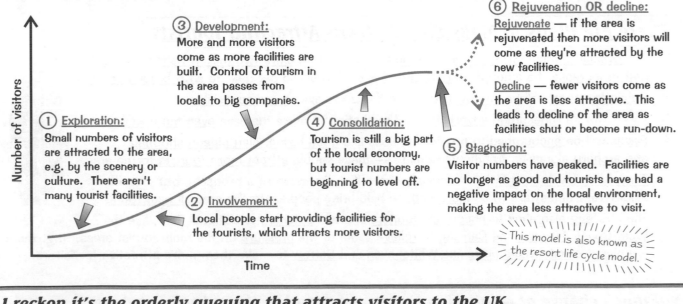

③ <u>Development:</u>
More and more visitors come as more facilities are built. Control of tourism in the area passes from locals to big companies.

⑥ <u>Rejuvenation OR decline:</u>
<u>Rejuvenate</u> — if the area is rejuvenated then more visitors will come as they're attracted by the new facilities.

<u>Decline</u> — fewer visitors come as the area is less attractive. This leads to decline of the area as facilities shut or become run-down.

① <u>Exploration:</u>
Small numbers of visitors are attracted to the area, e.g. by the scenery or culture. There aren't many tourist facilities.

④ <u>Consolidation:</u>
Tourism is still a big part of the local economy, but tourist numbers are beginning to level off.

⑤ <u>Stagnation:</u>
Visitor numbers have peaked. Facilities are no longer as good and tourists have had a negative impact on the local environment, making the area less attractive to visit.

② <u>Involvement:</u>
Local people start providing facilities for the tourists, which attracts more visitors.

This model is also known as the resort life cycle model.

Number of visitors (y-axis)

Time (x-axis)

I reckon it's the orderly queuing that attracts visitors to the UK...

The <u>tourist area life cycle model</u> applies to many <u>seaside resorts</u> in the UK. They were major tourism centres at the start of the 1900s but many <u>stagnated</u>, e.g. Morecambe, and <u>declined</u>. Some, e.g. Brighton, are now being <u>rejuvenated</u>.

UK Tourism — Case Study

The <u>Lake District National Park</u> has been a favourite with tourists since the early 1800s. Today, the huge number of visitors have to be <u>carefully managed</u> to preserve the natural beauty that's made it so popular.

The Lake District is a National Park in Cumbria

The <u>Lake District National Park</u> gets around <u>15 million</u> visitors <u>a year</u>.
There are <u>several reasons</u> it's <u>so popular</u>:

Lake District

1) Tourists come to enjoy the <u>scenery</u> — for example <u>large lakes</u> (e.g. <u>Windermere</u>) and <u>mountains</u> (e.g. <u>Scafell Pike</u>).

2) There are many <u>activities</u> available, e.g. <u>bird watching</u>, <u>walking</u>, <u>pony-trekking</u>, <u>boat rides</u>, <u>sailing</u> and <u>rock-climbing</u>.

3) There are also <u>cultural attractions</u>, e.g. the <u>Beatrix Potter</u> and <u>Wordsworth museums</u>.

Strategies are Needed to Cope with the Impact of Tourists

Tourists <u>cause traffic congestion</u>, <u>erode footpaths</u> and <u>drop litter</u>.
Here are a few strategies being carried out to <u>reduce</u> these problems:

1) **Coping with the extra traffic** — <u>public transport</u> in the area is <u>being improved</u> so people can <u>leave their cars at home</u>. There are also <u>campaigns</u> to <u>encourage</u> people to <u>use</u> the new services, e.g. the '<u>Give the driver a break</u>' campaign. This provides <u>leaflets</u> that show the <u>routes available</u> and offers <u>discounts</u> at cafes and on lake cruises for people presenting bus or train tickets.

2) **Coping with the erosion of footpaths** — solutions include encouraging visitors to <u>use less</u> <u>vulnerable</u> areas <u>instead</u>, '<u>resting</u>' popular routes by <u>changing the line of the paths</u>, and using more <u>hard-wearing materials</u> for paths. E.g. at <u>Tarn Hows</u>, severely eroded paths have been <u>covered</u> with <u>soil</u> and <u>reseeded</u>, and the <u>main route</u> has been <u>gravelled</u> to <u>protect it</u>.

3) **Protecting wildlife and farmland** — there are <u>signs</u> to remind visitors to <u>take their litter home</u> and <u>covered bins</u> are provided at the <u>most popular sites</u>. There have also been campaigns to <u>encourage</u> <u>visitors</u> to enjoy the countryside <u>responsibly</u>, e.g. by <u>closing gates</u> and <u>keeping dogs on a lead</u>.

There are Plans to Make Sure it Keeps Attracting Tourists

1) The <u>official tourism strategy for Cumbria</u> is to attract an <u>extra two million visitors</u> by <u>2018</u> and to <u>increase</u> the <u>amount tourists spend</u> from £1.1 billion per year to <u>£1.5 billion per year</u>.

2) <u>Public transport</u> will be <u>improved</u> to make the Lakes even <u>more accessible</u>.

3) There's to be <u>widespread advertising</u> and <u>marketing</u> to make the area <u>even more well known</u>.

4) <u>Farms</u> will be <u>encouraged</u> to provide services like <u>quad biking</u>, <u>clay pigeon shooting</u> and <u>archery</u> alongside traditional farming — these should <u>attract more tourists</u> to the area.

5) <u>Timeshare developments</u> (where people share the ownership of a property, but stay there at different times) are to be <u>increased</u>, to help bring people into the area <u>all year round</u>.

6) The strategy also aims to <u>encourage tourism</u> in areas <u>outside</u> the National Park, like the <u>West Coast</u>, <u>Furness</u> and <u>Carlisle</u>, to <u>relieve</u> some of the <u>pressure</u> on the main tourist areas. E.g. there are plans to <u>regenerate ports</u> like <u>Whitehaven</u> and <u>Barrow</u> to make them more attractive to visitors.

Not much chance of wandering lonely as a cloud in the Lakes nowadays...

What a lovely case study. Trust me, <u>visiting the Lakes</u> is a lot <u>more enjoyable</u> than learning about the area's <u>strategies</u> <u>for coping with tourism</u> — but the examiners <u>love</u> asking about this sort of thing, so you've really got to know it.

Mass Tourism

For me, 'mass tourism' conjures up images of pasty Brits swigging lager in Spanish coastal resorts, but there's a bit more to it than that. It can have a big impact on the areas the tourists flock to.

Mass Tourism is Basically Tourism on a Big Scale

Mass tourism is organised tourism for large numbers of people.
For example, visiting Spain on a package holiday would count as mass tourism.
But, holidays where people organise it themselves or small group tours don't count.

Mass Tourism has Both Positive and Negative Impacts

	POSITIVE	NEGATIVE
ECONOMIC IMPACTS	• It brings money into the local economy. • It creates jobs for local people, and increases the income of industries that supply tourism, e.g. farming.	• A lot of the profit made from tourism is kept by the large travel companies, rather than going to the local economy.
SOCIAL IMPACTS	• Lots of jobs means young people are more likely to stay in the area. • Improved roads, communications and infrastructure for tourists also benefit local people. • Income from tourism can be reinvested in local community projects.	• The tourism jobs available to locals are often badly paid and seasonal. • Traffic congestion caused by tourists can inconvenience local people. • The behaviour of some tourists can offend locals.
ENVIRONMENTAL IMPACTS	• Income from tourism can be reinvested in protecting the environment, e.g. to run National Parks or pay for conservation work.	• Transporting lots of people long distances releases lots of greenhouse gases that cause global warming. • Tourism can increase litter and cause pollution, e.g. increased sewage can cause river pollution. • Tourism can lead to the destruction of natural habitats, e.g. sightseeing boats can damage coral reefs.

There are Ways to Reduce the Negative Impacts of Mass Tourism

Here are a few examples:

1) Improving public transport encourages tourists to use it, which reduces congestion and pollution.

2) Limiting the number of people visiting sensitive environments, e.g. coral reefs, reduces damage.

3) Providing lots of bins helps to reduce litter.

The Importance of Tourism Needs to be Maintained

Areas that rely heavily on tourism need to make sure the tourists keep coming.
Here are a few ways they can do this:

1) Build new facilities or improve existing ones, e.g. build new hotels.

2) Reduce any tourist impacts that make the area less attractive, e.g. litter and traffic congestion.

3) Advertise and market the area to attract new tourists, e.g. use TV to advertise in other countries.

4) Improve transport infrastructure to make it quicker and easier to get to the area.

5) Offer new activities to attract tourists that don't normally go there.

6) Make it cheaper to visit, e.g. lower entrance fees to attractions.

O' I do like to live beside the seaside — or I did before that coach party turned up...

Pros and cons, that's what this page is about — the pros and cons of mass tourism obviously. Nothing to do with the criminal fraternity. You also need to know how tourism should be managed to stop those pesky cons from taking over.

Mass Tourism — Case Study

If watching lions devour a gazelle while on holiday is your bag, then Kenya is the place to go.

Kenya is a Popular Tourist Destination

Kenya is in East Africa. It gets over 700 000 visitors per year. There are a few reasons why people visit:

1) A fascinating tribal culture and lots of wildlife, including the 'big five' (rhino, lion, elephant, buffalo and leopard). Wildlife safaris are very popular.

2) A warm climate with sunshine all year round.

3) Beautiful scenery, including savannah, mountains, forests, beaches and coral reefs.

Tourism has Had a Big Impact on Kenya

	POSITIVE	NEGATIVE
ECONOMIC IMPACTS	• Tourism contributes 15% of the country's Gross National Product. • In 2003, around 219 000 people worked in the tourist industry.	• Only 15% of the money earned through tourism goes to locals. The rest goes to big companies.
SOCIAL IMPACTS	• The culture and customs of the native Maasai tribe are preserved because things like traditional dancing are often displayed for tourists.	• Some Maasai tribespeople were forced off their land to create National Parks for tourists. • Some Muslim people in Kenya are offended by the way female tourists dress.
ENVIRONMENTAL IMPACTS	• There are 23 National Parks in Kenya, e.g. Nairobi National Park. Tourists have to pay entry fees to get in. This money is used to maintain the National Parks, which help protect the environment and wildlife.	• Safari vehicles have destroyed vegetation and caused soil erosion. • Wild animals have been affected, e.g. cheetahs in the most heavily visited areas have changed their hunting behaviour to avoid the crowds. • Coral reefs in the Malindi Marine National Park have been damaged by tourist boats anchoring.

Kenya is Trying to Reduce the Negative Impacts of Tourism

1) Walking or horseback tours are being promoted over vehicle safaris, to preserve vegetation.

2) Alternative activities that are less damaging than safaris are also being encouraged, e.g. climbing and white water rafting.

Kenya is Also Trying to Maintain Tourism

1) Kenya's Tourist Board and Ministry of Tourism have launched an advertising campaign in Russia called 'Magical Kenya'.

2) Kenya Wildlife Service is planning to build airstrips in Ruma National Park and Mount Elgon National Park to make them more accessible for tourists. It also plans to spend £8 million improving roads, bridges and airstrips to improve accessibility.

3) Visa fees for adults were cut by 50% in 2009 to make it cheaper to visit the country. They were also scrapped for children under 16 to encourage more families to visit.

Shh! It's the lesser-spotted tourist in its natural habitat — the poolside bar...

A walking safari in Kenya sounds dodgy — lions eat people and elephants can be pretty stroppy when they want to be. It's just one way of reducing the impacts of tourism though (which reminds me — make sure you know the impacts too).

Tourism in Extreme Environments

Some people <u>aren't content</u> with a <u>week in the sun</u> or a <u>shopping spree</u> in New York —
they go on holiday to <u>extreme environments</u>, e.g. Antarctica, the Himalayas and the Sahara desert.

Extreme Environments are Becoming Popular with Tourists

There are many reasons why tourists are <u>attracted</u> to <u>extreme environments</u>:

1) They're ideal settings for <u>adventure holiday activities</u> like <u>jeep tours</u>,
<u>river rafting</u> and <u>trekking</u>.

2) Some people want something <u>different</u> and <u>exciting</u> to do on holiday,
which nobody else they know has done.

3) A lot of people enjoy an element of <u>risk</u>
and <u>danger</u> in their leisure time, which the <u>harsh conditions</u>
of an extreme environment can provide.

4) Some <u>wildlife</u> can <u>only be seen</u> in these areas,
e.g. polar bears can only be seen in the Arctic.

5) Some <u>scenery</u> can <u>only be seen</u> in extreme places too,
e.g. icebergs can only be seen in very cold environments.

There are also several reasons why tourism is <u>increasing</u> in <u>extreme environments</u>:

1) <u>Improvements in transport</u> have made it <u>quicker</u> and <u>easier</u> to <u>get to</u> some of these destinations.
For example, the Qinghai-Tibet railway that links China and Tibet (an extreme mountain environment)
opened in 2006. This increased tourism as Tibet was easier to get to.

Next year, I'll
do the packing.

2) People are keen to see places like <u>Antarctica</u> for themselves while they
<u>have the chance</u>, before the <u>ice melts</u> due to <u>global warming</u>.

3) Tourism to extreme environments is <u>quite expensive</u>, but people nowadays tend
to have <u>more disposable income</u> (spare cash), so <u>more people can afford to go</u>.

4) <u>Adventure holidays</u> are becoming <u>more popular</u> because of <u>TV programmes</u> and <u>advertising</u>.

Tourism in Extreme Environments can be Damaging

The <u>ecosystems</u> in extreme environments are usually <u>delicately balanced</u>, because it's so difficult for
life to survive in the <u>harsh conditions</u> there. The presence of tourists can <u>upset</u> this <u>fragile balance</u>
and cause <u>serious problems</u>. Here's an example of how tourism can <u>damage the environment</u> in
the <u>Himalayas</u>:

1) <u>Trees</u> are <u>cut down</u> to provide <u>fuel</u> for <u>trekkers</u>
and other tourists, leading to <u>deforestation</u>.

2) Deforestation <u>destroys habitats</u>.

3) Deforestation also means there are <u>fewer trees</u> to
<u>intercept rain</u>. So <u>more water reaches channels</u>
causing <u>flooding</u>.

4) <u>Tree roots</u> normally <u>hold the soil together</u>,
so deforestation also leads to <u>soil erosion</u>.
If soil is <u>washed into rivers</u> it <u>raises</u> the <u>river bed</u> so it
<u>can't hold as much water</u> — this can cause <u>flooding</u> too.

5) The sheer volume of tourists causes <u>footpath erosion</u>, which can lead to <u>landslides</u>.

6) <u>Toilets</u> are <u>poor</u> or <u>non-existent</u>, so <u>rivers</u> become <u>polluted</u> by <u>sewage</u>.

The inside of my fridge — now that's a pretty extreme environment...

In the exam they might ask you to suggest <u>reasons why</u> people go to extreme environments so you need to <u>learn them</u>
(answering 'because they're mad' won't cut it). Make sure you can <u>describe the impacts</u> of tourism for <u>one place</u> too.

Tourism in Extreme Environments — Case Study

Antarctica is the coldest place on Earth (it can get to minus 80 °C), making it an extreme environment. Despite this fact quite a few tourists don their thermal knickers and brave the cold every year.

The Antarctic is Becoming More Popular with Tourists

1) Antarctica is a continent at the Earth's South Pole. It covers an area of about 14 million km² and about 98% is covered with ice.

2) The number of tourists visiting Antarctica each year is rising, e.g. there were 7413 in the 1996/1997 season, but 46 000 in the 2007/2008 season.

3) Tourists are attracted by the stunning scenery (e.g. icebergs) and the wildlife (e.g. penguins and whales).

Tourism has Environmental Impacts in Antarctica

Antarctica is very cold and doesn't get much sunshine in winter so the land ecosystems are very fragile — it takes a long time for them to recover from damage. The sea ecosystem is also delicately balanced. This means that tourists can have a massive impact on the environment there:

> 1) Tourists can trample plants, disturb wildlife and drop litter.
>
> 2) There are fears that tourists could accidentally introduce non-native species or diseases that could wipe out existing species.
>
> 3) Spillage of fuel from ships is also a worry, especially after the sinking of the cruise ship, MS Explorer, in 2007. Fuel spills kill molluscs (e.g. mussels) and fish, as well as the birds that feed on them (e.g. penguins).

There are Measures in Place to Protect Antarctica

(1) The Antarctic Treaty is an international agreement that came into force in 1961 and has now been signed by 47 countries. The Treaty is designed to protect and conserve the area and its plant and animal life. In April 2009, the parties involved with the Antarctic Treaty agreed to introduce new limits on tourism in Antarctica — only ships with fewer than 500 passengers are allowed to land there and a maximum of 100 passengers are allowed on shore at a time.

(2) The International Association of Antarctica Tour Operators also has a separate Code of Conduct. The code is voluntary, but most operators in the area do stick to it. There are rules on:

1) Specially Protected Areas — these are off limits to tourists.

2) Wildlife — wildlife must not be disturbed when being observed. E.g. when whale watching, boats should approach animals slowly and keep their distance.

3) Litter — nothing can be left behind by tourists and there must be no smoking during shore landings (to reduce cigarette end litter).

4) Supervision — tourists must stay with their group and each group must have a qualified guide. This prevents people from entering no-go areas or disturbing wildlife.

5) Plant life — tourists must not walk on the fragile plant life.

6) Waste — sewage must be treated biologically and other waste stored on board the ships.

Please keep your voices down — the penguins are having an off day...

Time to get your 'case study hat' on — examiners expect you to be able to give details of tourism in a specific place. You need to know why people go there, how tourists affect the environment and how the area's being protected.

Ecotourism

As if UK tourism, mass tourism and extreme tourism weren't enough — it's time for ecotourism...

Ecotourism Doesn't Destroy the Environment

1) Ecotourism is tourism that doesn't harm the environment and benefits the local people.
2) Ecotourism involves: • Conservation — protecting and managing the environment.
   • Stewardship — taking responsibility for conserving the environment.
3) Ideally, conservation and stewardship should involve local people and local organisations, so that local people benefit from the tourists.
4) Ecotourism is usually a small-scale activity, with only small numbers of visitors going to an area at a time. This helps to keep the environmental impact of tourism low.
5) It often involves activities like wildlife viewing and walking.

Ecotourism Benefits the Environment, Economy and Local People

Environmental benefits:

1) Local people are encouraged to conserve the environment rather than use it for activities that can be damaging, e.g. logging or farming. This is because they can only earn money from ecotourism if the environment isn't damaged.
2) It reduces poaching and hunting of endangered species, since locals will benefit more from protecting these species for tourism than if they killed them.
3) Ecotourism projects try to reduce the use of fossil fuels, e.g. by using renewable energy sources and local food (which isn't transported as far so less fossil fuel is used). Using less fossil fuel is better for the environment as burning fossil fuels adds to global warming.
4) Waste that tourists create is disposed of carefully to prevent pollution.

Economic benefits:

1) Ecotourism creates jobs for local people (e.g. as guides or in tourist lodges), which helps the local economy grow.
2) Local people not directly employed in tourism can also make money by selling local crafts to visitors or supplying the tourist industry with goods, e.g. food.

Benefits for local people:

1) People have better and more stable incomes in ecotourism than in other jobs, e.g. farming.
2) Many ecotourism schemes fund community projects, e.g. schools, water tanks and health centres.

Ecotourism Helps the Sustainable Development of Areas

1) Sustainable development means improving the quality of life for people, but doing it in a way that doesn't stop people in the future getting what they need (by not damaging the environment or depleting resources).
2) Ecotourism helps areas to develop by increasing the quality of life for local people — the profits from ecotourism can be used to build schools or healthcare facilities.
3) The development is sustainable because it's done without damaging the environment — without ecotourism people may have to make a living to improve their lives by doing something that harms the environment, e.g. cutting down trees.

Not to be confused with egotourism — holidays on board a big golden yacht...

Ecotourism is sustainable because it's something that can continue into the future — the area remains unspoilt, so tourists will continue to come and enjoy it. It helps give local people a better quality of life too.

Ecotourism — Case Study

I know you're a bit sick of <u>case studies</u> by now, but this is the <u>last one</u> in the topic — scout's promise.

Tataquara Lodge is an Example of Ecotourism

1) <u>Tataquara Lodge</u> is on an island in the <u>Xingu River</u> in the <u>Brazilian</u> state of <u>Para</u>.

2) It's owned and operated by a <u>cooperative</u> of <u>six local tribes</u> of indigenous people.

3) The lodge has <u>15 rooms</u> and offers <u>activities</u> like <u>fishing</u>, <u>canoeing</u>, <u>wildlife viewing</u> and <u>forest walks</u>.

4) The surrounding rainforest is home to a rich variety of <u>wildlife</u>, including many species of bat and tropical birds. There are also some <u>endangered species</u> in the area, such as the <u>harpy eagle</u> and <u>giant river otter</u>.

Para is in the Amazon rainforest.

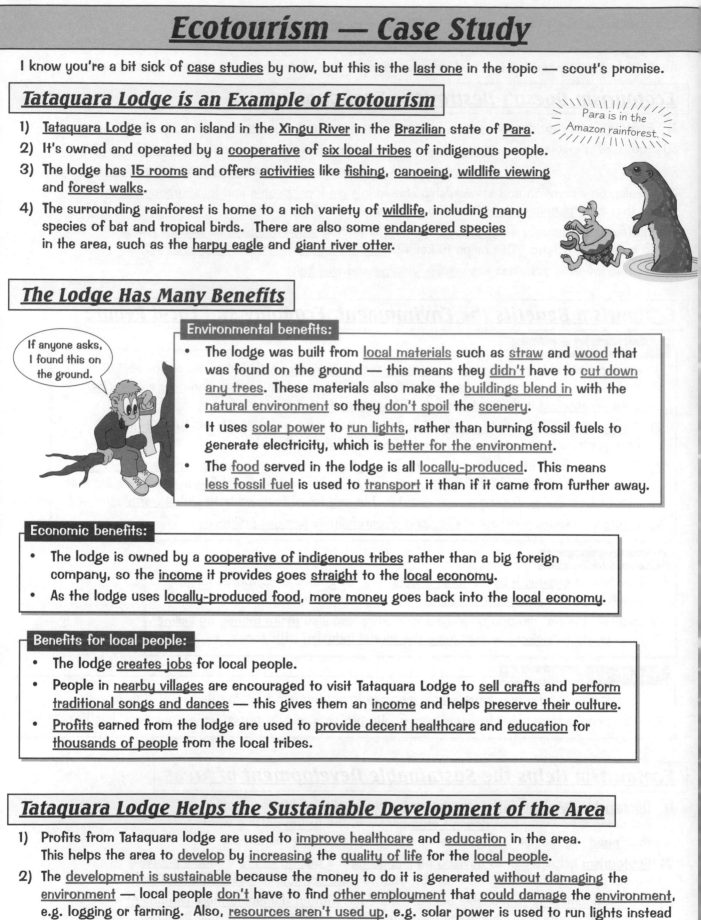

The Lodge Has Many Benefits

If anyone asks, I found this on the ground.

Environmental benefits:

• The lodge was built from <u>local materials</u> such as <u>straw</u> and <u>wood</u> that was found on the ground — this means they <u>didn't</u> have to <u>cut down any trees</u>. These materials also make the <u>buildings blend in</u> with the <u>natural environment</u> so they <u>don't spoil</u> the <u>scenery</u>.

• It uses <u>solar power</u> to <u>run lights</u>, rather than burning fossil fuels to generate electricity, which is <u>better for the environment</u>.

• The <u>food</u> served in the lodge is all <u>locally-produced</u>. This means <u>less fossil fuel</u> is used to <u>transport</u> it than if it came from further away.

Economic benefits:

• The lodge is owned by a <u>cooperative of indigenous tribes</u> rather than a big foreign company, so the <u>income</u> it provides goes <u>straight</u> to the <u>local economy</u>.

• As the lodge uses <u>locally-produced food</u>, <u>more money</u> goes back into the <u>local economy</u>.

Benefits for local people:

• The lodge <u>creates jobs</u> for local people.

• People in <u>nearby villages</u> are encouraged to visit Tataquara Lodge to <u>sell crafts</u> and <u>perform traditional songs and dances</u> — this gives them an <u>income</u> and helps <u>preserve their culture</u>.

• <u>Profits</u> earned from the lodge are used to <u>provide decent healthcare</u> and <u>education</u> for <u>thousands of people</u> from the local tribes.

Tataquara Lodge Helps the Sustainable Development of the Area

1) Profits from Tataquara lodge are used to <u>improve healthcare</u> and <u>education</u> in the area. This helps the area to <u>develop</u> by <u>increasing</u> the <u>quality of life</u> for the <u>local people</u>.

2) The <u>development is sustainable</u> because the money to do it is generated <u>without damaging</u> the <u>environment</u> — local people <u>don't</u> have to find <u>other employment</u> that <u>could damage</u> the <u>environment</u>, e.g. logging or farming. Also, <u>resources aren't used up</u>, e.g. solar power is used to run lights instead of fossil fuels, so <u>more resources</u> are <u>available</u> for <u>future generations</u>.

My local cooperative does a lovely strawberry cheesecake...

Tataquara may be a bit of a <u>mouthful</u> to say, but it's a <u>great example</u> of how ecotourism <u>benefits</u> the <u>environment</u> and the <u>local people</u>. Lovely stuff. Make sure you can <u>remember</u> a handful of <u>facts</u> about it for the exam.

Revision Summary for Tourism

It's that time again — just when you think you're all done and dusted with a topic, a page of questions is sprung on you. This lot are all about going on your holidays though, so they shouldn't be too much of a strain. Give them a try, and then if there are any you struggle with you can go back through the topic and pick up even more ideas for your next trip.

1) Give three reasons why global tourism is increasing.

2) Why do cities attract large numbers of tourists?

3) What attracts tourists to mountain areas?

4) Why is tourism important to the economies of many countries?

5) How much money does the tourist industry contribute to the UK's economy?

6) Give two factors that can decrease the number of tourists visiting the UK.

7) Describe the six stages in the tourist area life cycle model.

8) a) Name a UK National Park or UK coastal area.

   b) Describe what attracts tourists to the area.

   c) Describe the strategies used to reduce the impact of tourists.

   d) How does this area plan to keep attracting tourists in the future?

9) Define mass tourism.

10) List three positive effects and three negative effects of mass tourism.

11) Give three examples of how mass tourism can be managed to ensure that an area keeps its appeal.

12) a) Name a tropical tourist area that has mass tourism.

    b) Describe what makes this area such a popular tourist destination.

    c) Describe the negative impacts tourism has had on the area.

    d) What is being done to help reduce the negative impacts?

13) Give three reasons why tourism in extreme environments is increasing.

14) a) Name an extreme environment that is becoming popular with tourists.

    b) Describe the environmental impacts of tourism in the area.

    c) Describe the measures in place to limit the impact of tourism in the area.

15) Define ecotourism.

16) Explain one way that ecotourism can benefit the economy of a region.

17) Explain one way that ecotourism can benefit the environment of a region.

18) How can ecotourism help benefit local people?

19) a) What is sustainable development?

    b) Explain how ecotourism helps the sustainable development of areas.

20) a) Give an example of a successful ecotourism project.

    b) Explain how this project benefits the local environment, the local economy and the local people.

Unit 3 — Local Fieldwork Investigation

Local Fieldwork Investigation — Overview

Congratulations on making it to the second to last section of the book. Only two more sections to go and you're home and dry. This section covers the exciting world of the <u>local fieldwork investigation</u>, yey...

The Investigation Involves Fieldwork and a Report

1) Geography is about <u>where</u> things happen, and explaining <u>how</u> and <u>why</u> they happen, e.g. where coastal erosion occurs, and how and why it happens.

2) Geographers <u>ask a question</u>, carry out <u>fieldwork</u>, <u>analyse the data</u> collected and use the <u>results</u> to <u>answer</u> the original question. And you get to do all this yourself during the <u>local fieldwork investigation</u>.

3) It's worth <u>25%</u> of your final mark and involves collecting <u>primary data</u> (data <u>you collect</u> during fieldwork) and secondary data (data <u>other people</u> have collected).

4) Then you need to write a <u>2000 word report</u> all about it. The report can be <u>hand written</u> or <u>typed</u> on a computer. You need to <u>present</u> all the data you've collected, e.g. using annotated maps, graphs, etc.

5) Here's a simple <u>flow chart</u> to show you what you'll need to do and <u>roughly</u> how much <u>time</u> you should spend on each part:

> A hypothesis is a statement that you can investigate to see if it's true or false, e.g. 'rural-urban migration is caused by a shortage of services in rural communities'.

---

1. Pick a <u>task statement</u> and make up a <u>question</u> or <u>hypothesis</u>.

Pick <u>one task statement</u> from the 11 task statements provided by the exam board. <u>Ask your teacher</u> if you're unsure which task you're completing. Then <u>ask a question</u> or form a <u>hypothesis</u> that's relevant to your task statement. E.g. task — investigate the impact of public transport networks, hypothesis — the tram system in Bristelton has had a positive impact, OR question — has the tram system in Bristelton had a positive or negative impact?

---

2. Carry out some <u>preparation</u> — 4 hours.

<u>Research</u> the <u>area</u> you're investigating, the <u>processes</u> and <u>concepts</u> involved and collect any <u>secondary data</u> that might be useful, e.g. statistics. <u>Plan how</u> you're going to <u>collect your data</u> and <u>what you'll do with it</u>, e.g. write a <u>questionnaire</u> and decide <u>where</u> you'll conduct it. Think about <u>how</u> you'll <u>present</u> the <u>data</u>.

Page 133

---

3. Carry out the <u>fieldwork</u> and <u>collect your data</u> — no time limit.

---

4. <u>Write up</u> the <u>first part</u> of your <u>report</u> — 10 hours, 800 words.

   a) <u>Introduction</u> — introduce the <u>task statement</u> and your <u>question</u> or <u>hypothesis</u>. <u>Introduce</u> the <u>area</u> and any relevant <u>geographical processes</u> or <u>concepts</u>.

   b) <u>Methodology</u> — <u>describe</u> the methods you used and justify <u>why</u> you used them.

   c) <u>Data processing</u> — <u>organise</u> and <u>manipulate</u> your fieldwork data.

   d) <u>Data presentation</u> — <u>present</u> your fieldwork data using graphs, tables, etc.

Page 133 & 134

---

5. <u>Write up</u> the <u>second part</u> of your <u>report</u> — 6 hours, 1200 words.

   a) <u>Analyse and interpret evidence</u> — <u>describe</u> your data and <u>explain</u> what it <u>shows</u>.

   b) <u>Conclusion</u> — give <u>reasons</u> for the results and use your data to <u>answer</u> your question or say whether it <u>supports</u> your hypothesis or not.

   c) <u>Evaluation</u> — discuss how <u>good</u> the data collection methods you used were, whether this affects the <u>accuracy</u> of your <u>results</u> and how <u>valid</u> your <u>conclusion</u> is.

Page 135 & 136

Hippos like wallowing — my hippo-thesis...

Your local fieldwork investigation is worth <u>60 marks</u> and makes up <u>25%</u> of your total grade. How you get these marks is explained in the <u>marking criteria</u> — ask your teacher for a copy so you can see what a good report should contain.

Preparation, Introduction and Methodology

Now you've got a general idea what the investigation involves, let's get down to the details. There's nowhere better to start than <u>preparation</u>, in fact here's some preparation I prepared earlier... (boom boom)

Good Preparation will Help You Later On

Here are a few of the things you could <u>do</u> during the <u>preparation time</u>:

Collect secondary data —

Look in <u>newspapers</u>, <u>books</u> and on <u>websites</u>. Keep a <u>record</u> of <u>where</u> the data comes from so you can create a <u>bibliography</u>.

Research the area you're investigating —

Collect <u>background information</u> on the area you're studying (e.g. photos and maps). Study the <u>processes</u> or <u>concepts</u> that gave the area its <u>geographical features</u>.

Plan your fieldwork —

Plan what <u>methods</u> you'll use and what <u>equipment</u> you'll need. Think about <u>how much</u> data you'll collect — generally the <u>more</u> you collect the <u>more reliable</u> your <u>results</u> will be. Also think about how much <u>time</u> it'll take and <u>where</u> to collect data. And don't forget about <u>health and safety</u>.

You can also ask your <u>teacher</u> for <u>advice</u> on things like your <u>question</u>, <u>techniques</u> and <u>report layout</u>. Keep your <u>notes</u> in a <u>folder</u> and hand it in to your teacher at the end of each lesson. You can only use <u>your</u> notes in the write-up sessions so make sure they're <u>good</u> and <u>thorough</u>.

First part of the Report — a) The Introduction

1) <u>Introduce</u> the <u>task statement</u> and the <u>question</u> or <u>hypothesis</u> you're investigating. <u>Explain why</u> it could be <u>important</u> to other people. E.g. investigating the impact of a <u>tram system</u> on city centre <u>traffic</u> could be used to see if a <u>similar scheme</u> would be a good way to <u>reduce</u> traffic in another city.

2) Introduce the <u>study area</u> and its <u>location</u> <u>in detail</u> and <u>explain why</u> you're investigating <u>that area</u>. Include things like <u>annotated photos</u> and <u>maps</u> with <u>titles</u>, <u>grid references</u>, <u>scales</u> and <u>keys</u>. <u>Mark on</u> the maps and photos <u>where</u> you <u>collected data</u>.

3) <u>Describe</u> and <u>explain</u> the <u>geographical processes and concepts</u> you're investigating — use this book to help. E.g. if you're looking at <u>rural-urban migration</u> explain <u>push and pull factors</u> and give some examples. If you're investigating <u>coastal erosion management</u> <u>describe</u> what coastal erosion <u>is</u> and <u>how</u> it occurs.

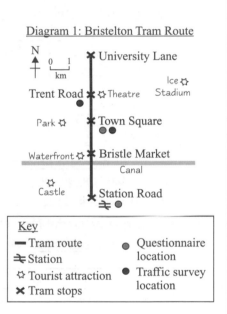

Diagram 1: Bristelton Tram Route

First part of the Report — b) The Methodology

You need to <u>describe</u> all the data collection methods you used, what they <u>involved</u> and <u>why</u> you used them. One way of showing this information is in a <u>table</u>:

Method	Type of data	Description	Size	Where and when	Why
Questionnaire	Primary	Describe the method, e.g. say what you did and refer them to the blank questionnaire you've included.	Say how many questions were asked, how many people you asked and at what sites.	Describe where you asked people the questions (refer to any maps that show where) and when you asked them (time, day and date).	Explain why you chose a questionnaire and why you chose those questions, locations and numbers of people.

You can also draw diagrams to show how you used equipment.

There's method to my madness — you can read all about it in my table...

You don't need to use a <u>table</u> to describe and explain your methods. It's just a handy way to do it that means you shouldn't miss anything out — if you did miss anything, there'd be an <u>empty box</u>, which you'd notice. Super.

Data Processing and Presentation

You've made it to the fun part now — <u>processing</u> and <u>presenting</u> your data. It's graphs galore.
And as it's geography, I'd bet my good arm (left if you're asking) that there'll be some colouring-in too.

First part of the Report — c) Data Processing

Now you've got lots of <u>primary</u> and <u>secondary data</u> lying around, you need to <u>process</u> it.

1) <u>Data processing</u> can mean <u>organising</u> data to make it easier to <u>analyse</u> and <u>understand</u>. E.g. it's no good putting <u>every</u> questionnaire answered in your report, it'd be <u>too long</u> and hard to analyse. Instead, you could make a table that gives a <u>summary</u> of the answers. Don't forget to include a <u>copy</u> of the blank questionnaire though, so you can <u>refer</u> to the questions you asked:

Q1	Why have you come into Bristelton today?
Q2	How far have you travelled to come to Bristelton today?
Q3	How did you get to Bristelton today?

Question 1 answers:

Location	Shopping	Tourism	Work	Football	Other
Town Square	15	10	6	16	3
Station Road	10	6	3	25	6

2) You can gain marks by <u>manipulating</u> your data before you present it too, e.g. you could calculate <u>percentages</u>, or <u>averages</u> (<u>means</u> and <u>modes</u>). The data can then be presented in a way that's <u>easier to understand</u>, e.g. a <u>pie chart</u>.

3) You can also use <u>equations</u> to manipulate your data before presenting it, e.g. the <u>discharge</u> of a river = <u>cross-sectional area</u> × average <u>velocity</u>.

Question 3 answers:

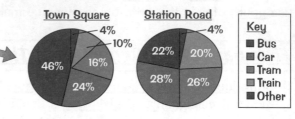

Town Square — 4%, 10%, 16%, 46%, 24%
Station Road — 4%, 20%, 22%, 26%, 28%

Key
■ Bus
■ Car
■ Tram
■ Train
■ Other

First part of the Report — d) Data Presentation

1) Use lots of <u>different presentation techniques</u> in your report.
For example, <u>tables</u>, <u>pie charts</u>, <u>graphs</u>, <u>diagrams</u>, <u>maps</u>, <u>annotated sketches</u> and <u>photographs</u>.

2) Make sure you include clear <u>titles</u>, <u>scales</u>, <u>units</u> and <u>keys</u> in your data presentation.

3) At least <u>one</u> presentation technique should be done on a <u>computer</u>, e.g. make some graphs in Excel®.

4) Check out the <u>examples</u> below and pages 140-143 for other ways of <u>presenting data</u>:

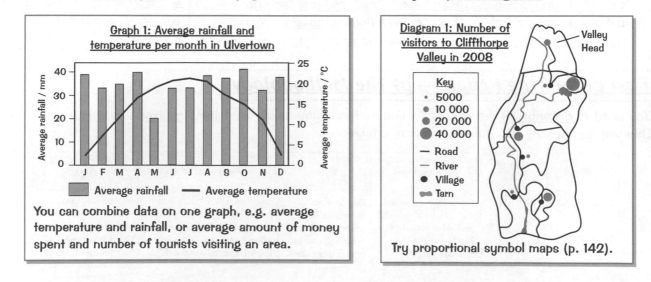

Graph 1: Average rainfall and temperature per month in Ulvertown

You can combine data on one graph, e.g. average temperature and rainfall, or average amount of money spent and number of tourists visiting an area.

Diagram 1: Number of visitors to Cliffthorpe Valley in 2008

Key
· 5000
• 10 000
⬤ 20 000
⬤ 40 000
— Road
— River
• Village
⬗ Tarn

Valley Head

Try proportional symbol maps (p. 142).

Data presentation — I find wrapping paper and a pretty bow help...

There are loads of different ways to present your data so don't just stick in <u>bar charts</u> and <u>tables</u>. Try some new and exciting techniques like <u>choropleth maps</u> or <u>flow line maps</u> — look at page 143 if you don't know what they are.

Analysis and Conclusions

Now we're onto the slightly trickier parts — analysis and conclusions. For this part of the report you won't be allowed to talk to your friends but you can look at the notes in your folder (I told you to make them good). You can ask your teacher questions too, but otherwise this bit is done under exam conditions.

Second part of the Report — a) Data Analysis and Interpretation

1) Describe what your data shows — describe any patterns and correlations (see pages 139 and 141) and look for any anomalies (odd results). Make comparisons between different sets of data, use specific points from your data and reference what graph, table etc. you're talking about.

2) Explain what your data shows — explain why there are patterns and why different data sets are linked together. Use your geographical background information to help you explain why you got your results and remember to use geographical terms.

   For example:

   > 38% of people who visited Cliffthorpe Valley in 2008 visited the tarn — 40 000 people (see Diagram 1). The tarn area may attract visitors due to its beauty and services such as a free car park, café and tourist information centre (see Leaflet 1). However, the largest amount of litter was found at the valley head (see Graph 1), which was the fourth most popular attraction (9.5% of visitors). There are fewer bins at the valley head, and more people tend to picnic there (see Table 1), which could be why there's more litter.

Second part of the Report — b) Conclusion

A conclusion is a summary of what you found out in relation to your original question or hypothesis. You need to:

Be careful when drawing conclusions. Some results show a link or correlation, but that doesn't mean that one thing causes another.

1) Write a summary of your results — look at all your data and describe generally what it shows.

2) If you're investigating a question — give an answer to the question and explain why.
   If you're investigating a hypothesis — say whether your data supports the hypothesis or not and explain why.

3) Explain how your conclusion fits into the wider geographical world — think about how your conclusion and results could be used by other people or in further investigations.

   For example:

   > I believe that my results support my original hypothesis that the tram system has had a positive impact on Bristelton. The results of my questionnaire showed that local people agreed that it has had a positive impact on the town. At the weekend, the number of people entering the town by car appears to have decreased, suggesting a positive environmental impact of the system. The number of people using the bus at the weekend has stayed similar to levels before the tram, suggesting it has encouraged more public transport use, rather than just taking passengers away from the bus. I think my investigation would be useful for other towns considering implementing a tram system. For example, negative comments in questionnaires like 'the trams are too noisy' could be considered when choosing the make of the new trams.

You vill answer all my questions — ve have vays of making you talk...

You get marks for punctuation, grammar and spelling, so make sure you've checked all these. If it's word processed you can use the spellchecker. Don't forget to include plenty of geographical terms where you can too.

Evaluation and Checklist

The last page of the section, yippe-ky-yay... The final part of the report involves <u>evaluating</u> what you did, and no, just putting you were brilliant and the investigation had no faults won't cut it.

Second part of the Report — c) Evaluation

Evaluation is all about <u>self assessment</u> — looking back at how <u>good or bad</u> your study was. You need to:

1) Identify any <u>problems</u> with the <u>methods</u> you used and suggest how they could be <u>improved</u>. Think about things like the <u>size</u> of your <u>data sets</u>, if any <u>bias</u> slipped in and if <u>other methods</u> would have been <u>more appropriate</u> or <u>more effective</u>.

2) Describe how <u>accurate</u> your results are and <u>link</u> this to the methods you used — say whether any <u>errors</u> in the methods affected the results.

3) Comment on the <u>validity</u> of your <u>conclusion</u>. You need to talk about how <u>problems</u> with the methods and the <u>accuracy</u> of the results affect the <u>validity</u> of the conclusion. Problems with methods lead to <u>less</u> reliable and accurate results, which affects the validity of the conclusion.

> <u>Accurate</u> results are <u>as near</u> as possible to the <u>true answer</u> — they have <u>few errors</u>.
> <u>Reliable</u> means that data can be <u>reproduced</u>.
> <u>Valid</u> means that the data <u>answers</u> the <u>original question</u> and is <u>reliable</u>.

Example:

> One <u>problem</u> with my questionnaire was that it was carried out on a <u>Saturday</u> when there was a <u>football</u> match on. This meant that some people questioned had come into Bristelton for a <u>specific reason</u> and may <u>not usually</u> come into the town. This may have affected the <u>results</u> as <u>fewer local people</u> or people that regularly come into Bristelton were consulted. So my conclusion, that <u>local people</u> agreed the tram system had a positive impact, <u>may not</u> reflect the majority view of <u>local residents</u>. Carrying out the <u>same</u> <u>questionnaire</u> on a weekday when local people are travelling in for work may produce <u>more accurate and reliable</u> results and so a more <u>valid conclusion</u>.

Check You've Included Everything with this Handy Checklist

I'm pretty confident you've got your report all sorted now. But if you want to make absolutely sure you've not missed out any biggies, here's a cheeky little <u>checklist</u> you can use.
<u>Tick</u> the box once you've done each thing:

1)	Introduced your question or hypothesis.	☐
2)	Introduced the study area in detail.	☐
3)	Described the geographical processes or concepts.	☐
4)	Described what you did, when, where and why you did it for each method.	☐
5)	Processed your data and included many different presentation techniques (using a computer for some of them).	☐
6)	Included titles, keys, scales and units when presenting data. The diagrams have been numbered so you can refer to them.	☐
7)	Described what your data shows.	☐
8)	Explained what your data shows.	☐
9)	Given a summary of your results in your conclusion.	☐

10)	Checked your conclusion answers your original question or supports your hypothesis.	☐
11)	Explained if your conclusion can be used by other people.	☐
12)	Identified any problems with your methods.	☐
13)	Described any possible improvements.	☐
14)	Described whether the results are accurate and the conclusion is valid.	☐
15)	Included a title page, contents table or page, page numbers and bibliography.	☐
16)	Checked your spelling, punctuation and grammar.	☐
17)	Used geographical terms.	☐

Present for the wonderful people at CGP for making this book — check...

And that my friends (as they say in show business) is that. OK, so no one says that in show business, but someone, somewhere, says it. Anyway, you've made it to the <u>end of the section</u>, well done. Feel free to feel smug.

Answering Questions

This section is filled with lots of lovely <u>techniques</u> and <u>skills</u> that you need for your <u>exam</u>. It's no good learning the <u>content</u> of this book if you don't bother learning the skills you need to pass your exam too. First up, answering questions properly...

Make Sure you Read the Question Properly

It's dead easy to <u>misread</u> the question and spend five minutes writing about the <u>wrong thing</u>. Four simple tips can help you <u>avoid</u> this:

1) Figure out if it's a <u>case study question</u> — if the question wording includes 'using <u>named</u> <u>examples</u>' or 'with reference to one <u>named</u> area' you need to include a case study.

2) <u>Underline</u> the <u>command words</u> in the question (the ones that tell you <u>what to do</u>):

> Answers to questions with 'explain' in them often include the word '<u>because</u>' (or '<u>due to</u>').

> When writing about differences, '<u>whereas</u>' is a good word to use in your answers, e.g. 'the Richter scale measures the energy released by an earthquake whereas the Mercalli scale measures the effects'.

Command word	Means write about...
Describe	what it's <u>like</u>
Explain	<u>why</u> it's like that (i.e. give <u>reasons</u>)
Compare	the <u>similarities</u> AND <u>differences</u>
Contrast	the <u>differences</u>
Suggest why	give <u>reasons</u> for

> If a question asks you to describe a <u>pattern</u> (e.g. from a map or graph), make sure you identify the <u>general pattern</u>, then refer to any <u>anomalies</u> (things that <u>don't</u> fit the general pattern).
>
> E.g. to answer 'describe the global distribution of volcanoes', first say that they're mostly on plate margins, <u>then</u> mention that a few aren't (e.g. in Hawaii).

3) <u>Underline</u> the <u>key words</u> (the ones that tell you what it's <u>about</u>), e.g. volcanoes, tourism, immigrants, rural-urban fringe, population pyramid.

4) <u>Re-read</u> the <u>question</u> and your <u>answer</u> when you've <u>finished</u>, just to check that what you've written really does <u>answer</u> the question being asked. A common mistake is to <u>miss a bit out</u> — like when questions say 'use <u>data</u> from the graph in your answer' or 'use <u>evidence</u> from the map'.

Case Study Questions are Level Marked

Case study questions are often worth <u>8 marks</u> and are <u>level marked</u>, which means you need to do these <u>things</u> to get the <u>top level</u> (3) and a <u>high mark</u>:

1) <u>Read</u> the question properly and figure out a <u>structure</u> before you start.
   Your answer needs to be well <u>organised</u> and <u>structured</u>, and written in a <u>logical</u> way.

2) Include <u>specialist terms</u> (geographical words), e.g. fold mountains, backwash, longshore drift.

3) Include plenty of <u>relevant details</u>:

   - This includes things like <u>names</u>, <u>dates</u>, <u>statistics</u>, names of <u>organisations</u> or <u>companies</u>.
   - Don't forget that they need to be <u>relevant</u> though — it's no good including the exact number of people killed in a flood when the question is about the <u>causes</u> of a flood.

4) Your answer should also be <u>legible</u> (you won't get many marks if the examiner <u>can't read</u> it). You should use correct <u>grammar</u>, and everything should be <u>spelt correctly</u> (double-check jazzy geography words).

Describe the similarities and differences between compare and contrast...

It may all seem a bit simple to you, but it's really important to understand what you're being <u>asked to do</u>. This can be tricky — sometimes the <u>differences</u> between the meanings of the command words are quite <u>subtle</u>, so get learnin'.

Labelling and Comparing

These next few pages give you some advice on what to do for <u>specific types</u> of questions.
Some of these skills will be helpful for your <u>fieldwork investigation</u> too (see pages 132-136).

You Might have to Label Photos, Diagrams or Maps

If you're asked to <u>label</u> something:

1) Figure out from the question what the <u>labels should do</u>, e.g. <u>describe</u> the <u>effects</u> of an earthquake, label the <u>characteristics</u> of a waterfall, <u>describe</u> the <u>coastal defences</u>, etc.

2) Add <u>at least</u> as many labels as there are <u>marks</u>.

3) When <u>describing</u> the <u>features</u> talk about things like the <u>size</u>, <u>shape</u> and <u>relief</u>.
Make sure you use the correct <u>geographical names</u> of any features,
e.g. tor, escarpment, meander.

Q: Label the <u>characteristics</u> of this coastline.

A:

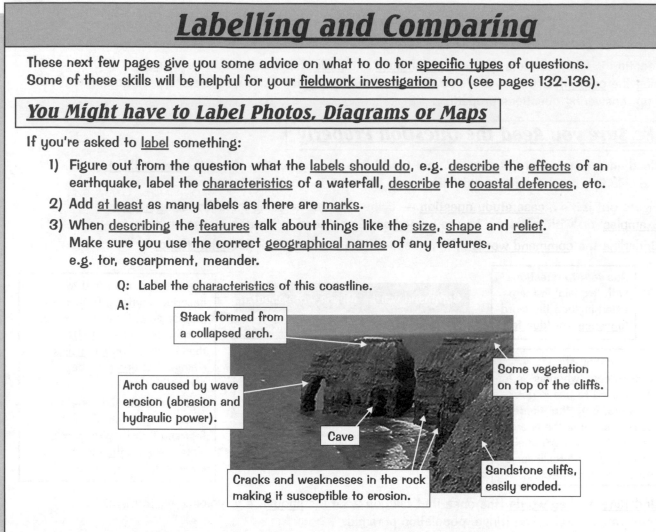

- Stack formed from a collapsed arch.
- Some vegetation on top of the cliffs.
- Arch caused by wave erosion (abrasion and hydraulic power).
- Cave
- Cracks and weaknesses in the rock making it susceptible to erosion.
- Sandstone cliffs, easily eroded.

Look at Shapes When You Compare Plans and Photos

You might be given <u>two items</u>, like a <u>plan</u> and an <u>aerial photograph</u>, and be asked to use them <u>together</u> to answer some questions. Plans and aerial photos are a bit like <u>maps</u> — they show places from <u>above</u>. Here are some <u>tips</u> for questions that use plans and photos:

1) The plan and photo might <u>not</u> be the <u>same way up</u>.

2) Work out how the photo <u>matches</u> the plan — look for the main <u>features</u> on the plan like a <u>lake</u>, a <u>big road</u> or something with an <u>interesting shape</u>, and find them on the photo.

3) Look at what's <u>different</u> between the plan and the photo and think about <u>why</u> it might be different.

Q: Look at the <u>development plan</u> for Crystal Bay (2000) and the <u>photo</u> taken <u>after development</u> in 2009.

   a) <u>Name</u> the area labelled A in the photo.

   b) Give one <u>difference</u> you can see between the photo and the plan.

A: a) Madeleine Park

   b) The roads have been built in slightly different areas. There's a small harbour area in front of the apartments.

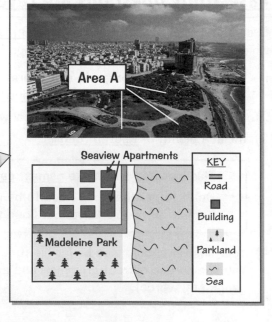

Area A

Seaview Apartments

Madeleine Park

KEY
Road
Building
Parkland
Sea

It isn't only fashionistas that are interested in labels...

You might have to use <u>plans</u> or <u>photos</u> in your exam to answer all sorts of questions — take your time and <u>read</u> the question carefully so you know exactly what you should be doing. Up next <u>maps and graphs</u>, whoop whoop.

Describing Maps and Graphs

One thing we geographers know how to do well is colouring in <u>maps</u> and <u>graphs</u>. You might get maps and graphs in the exam, but you'll probably have to <u>describe what they show</u>, rather than break out the crayons.

Describing Distributions on Maps — Describe the Pattern

1) In your exam you could get questions like, 'use the map to <u>describe</u> the <u>distribution</u> of volcanoes' and '<u>explain</u> the <u>distribution</u> of deforestation'.

2) Describe the <u>general pattern</u> and any <u>anomalies</u> (things that <u>don't fit</u> the general pattern).

3) Make <u>at least</u> as many <u>points</u> as there are <u>marks</u> and use <u>names</u> of places and <u>figures</u> if they're given.

4) If you're asked to give a <u>reason</u> or <u>explain</u>, you need to describe the <u>distribution first</u>.

Figure 1 — Population density of the UK

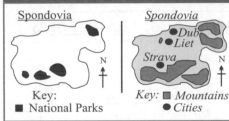

Key
- 600 to 5000 persons per km^2
- 400 to 599 persons per km^2
- 200 to 399 persons per km^2
- 0 to 199 persons per km^2

N

Q: Use Figure 1 to explain the pattern of population density in the UK.

A: The <u>London area</u> has a <u>very high</u> population density (<u>600 to 5000</u> per km²). There are also areas of <u>high</u> population density (<u>400 to 599</u> per km²) in the <u>south east</u> and <u>west</u> of England. These areas include <u>major cities</u> (e.g. Birmingham and Manchester). More people live in and around cities because there are <u>better services</u> and <u>more job opportunities</u> than in rural areas. <u>Scotland</u> and <u>Wales</u> have the <u>lowest</u> population density in the UK (<u>less than 199</u> per km²)...

You could be given two maps to use for one question — link information from the two maps together.

Describing Locations on Maps — Include Details

1) In your exam you could get a question like, 'suggest a <u>reason</u> for the <u>location</u> of the settlement'.

2) When you're asked about the <u>location</u> of something say <u>where</u> it is, what it's <u>near</u> and use <u>compass points</u>.

3) If you're asked to give a <u>reason</u> or <u>explain</u>, you need to describe the <u>location first</u>.

Q: Use the maps to describe the location of the National Parks.

Spondovia

Spondovia
- Dub
- Liet
- Strava

N

Key:
■ National Parks

Key: ■ Mountains
● Cities

A: The National Parks are found in the <u>south west</u> and <u>north east</u> of Spondovia. They are all located in <u>mountainous</u> areas. Three of the parks are located near to the city of <u>Strava</u>.

Describing what Graphs Show — Include Figures from the Graph

When <u>describing</u> graphs make sure you mention:

1) The general pattern — when it's <u>going up</u> and <u>down</u>, and any <u>peaks</u> (highest bits) and <u>troughs</u> (lowest bits).

2) Any <u>anomalies</u> (odd results).

3) Specific <u>data points</u>.

Q: Use the graph to describe population change in Cheeseham.

A: The population halved between 1950 and 1960 from 40 thousand people to 20 thousand people. It then increased to 100 thousand by 1980, before falling slightly and staying steady at 90 thousand from 1990 to 2000.

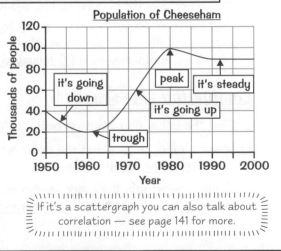

Population of Cheeseham

it's going down

peak | it's steady

it's going up

trough

If it's a scattergraph you can also talk about correlation — see page 141 for more.

Misfit data points with hats and fishing rods — a-gnome-alies...

Get it... a-gnome-alies.. I quite like gnomes, but then I also like studying <u>maps</u> and <u>graphs</u> just for kicks. I really should get out more, maybe I could go and visit the good people of Cheeseham... mmm... cheese.

Charts and Graphs

The next four pages are filled with lots of __different__ types of __charts__, __graphs__ and __maps__. There are two __important__ things to learn — NUMBER ONE: how to __interpret__ them (read them), and NUMBER TWO: how to __construct__ and __complete__ them (fill them in). You might have to do it in the __exam__ so pay attention.

Bar Charts — Draw the Bars Straight and Neat

① Reading Bar Charts

1) Read along the __bottom__ to find the __bar__ you want.

2) To find out the __value__ of a bar in a __normal__ bar chart — go from the __top__ of the bar __across__ to the __scale__, and __read off__ the number.

3) To find out the __value__ of __part__ of the bar in a __divided__ bar chart — find the __number at the top__ of the part of the bar you're interested in, and __take away__ the __number at the bottom__ of it.

Q: How many barrels of oil did Oxo oil produce per day in 2008?

A: 500 thousand − 350 thousand = __150 thousand barrels__ per day

Oil production

(chart: 2007 / 2008, Line across from 350, Thousands of barrels per day vs Company: Oxo oil, Gnoxo Ltd., Froxo Inc.)

② Completing Bar Charts

1) First find the number you want on the __vertical scale__.

2) Then __trace__ a line across to where you want the __top__ of the bar to be with a __ruler__.

3) Draw in a bar of the __right size__ using a __ruler__.

Q: Complete the chart to show that Froxo Inc. produced 200 thousand barrels of oil per day in 2008.

A: 150 thousand (2007) + 200 thousand = __350 thousand barrels__. So draw the bar up to this point.

Line Graphs — the Points are Joined by Lines

① Reading Line Graphs

1) Read along the __correct scale__ to find the __value__ you want, e.g. 20 thousand tonnes or 1920.

2) Read __across__ or __up__ to the line you want, then read the value off the __other__ scale.

Q: How much coal did New Wales Ltd. produce in 1900?

A: Find 1900 on the bottom scale, go up to the red line, read across, and it's 20 on the scale. The scale's in thousands of tonnes, so the answer is __20 thousand tonnes__.

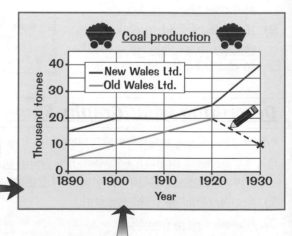

Coal production

(graph: New Wales Ltd. / Old Wales Ltd., Thousand tonnes vs Year: 1890–1930)

② Completing Line Graphs

1) Find the value you want on __both scales__.

2) Make a __mark__ (e.g. ×) at the point where the __two values meet__ on the graph.

3) Using a __ruler__, __join__ the __mark__ you've made to the __line__ that it should be __connected to__.

Q: Complete the graph to show that Old Wales Ltd. produced 10 thousand tonnes of coal in 1930.

A: Find 1930 on the bottom scale, and 10 thousand tonnes on the vertical scale. Make a mark __where they meet__, then join it to the __blue__ line __with a ruler__.

The top forty for sheep — the baaaaaaaaaaaaa chart...

Something to watch out for with __bar charts__ and __line graphs__ is reading the __scale__ — check how much each division is __worth__ before reading them or completing them. It's easy to think they're always worth one each, but sadly not.

Charts and Graphs

'More <u>charts</u> and <u>graphs</u>' I hear you cry — well OK, your weird wishes are my command.

Scatter Graphs Show Relationships

<u>Scatter graphs</u> tell you how <u>closely related</u> two things are, e.g. altitude and air temperature. The fancy word for this is <u>correlation</u>. <u>Strong</u> correlation means the two things are <u>closely</u> related to each other. <u>Weak</u> correlation means they're <u>not very</u> closely related. The <u>line of best fit</u> is a line that goes roughly through the <u>middle</u> of the scatter of points and tells you about what <u>type</u> of correlation there is. Data can show <u>three</u> types of correlation:

1) <u>Positive</u> — as one thing <u>increases</u> the other <u>increases</u>.
2) <u>Negative</u> — as one thing <u>increases</u> the other <u>decreases</u>.
3) <u>None</u> — there's <u>no relationship</u> between the two things.

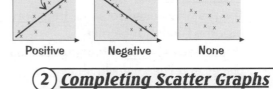
Positive   Negative   None

1 Reading Scatter Graphs

1) If you're asked to <u>describe</u> the <u>relationship</u>, look at the <u>slope</u> of the graph, e.g. if the line's moving <u>upwards</u> to the <u>right</u> it's a <u>positive correlation</u>. You also need to look at how <u>close</u> the points are to the <u>line of best fit</u> — the <u>closer</u> they are the <u>stronger</u> the correlation.

2) If you're asked to read off a <u>specific point</u>, just follow the <u>rules</u> for a <u>line graph</u> (see previous page).

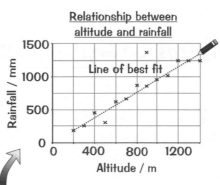

Relationship between altitude and rainfall

Q: Describe the relationship shown by the scatter graph.
A: Altitude and rainfall show a strong, positive correlation — as altitude increases, so does the amount of rainfall.

2 Completing Scatter Graphs

1) You could be asked to <u>draw</u> a <u>line of best fit</u> — just draw it roughly through the <u>middle</u> of the scatter of points.

2) If you're asked to <u>add a point</u> — just follow the <u>rules</u> for adding a point to a <u>line graph</u> (see previous page).

Pie Charts Show Amounts or Percentages

The important thing to remember with pie charts is that <u>the whole pie = 360°</u>.

1 Reading Pie Charts

1) To work out the <u>%</u> for a wedge of the pie, use a <u>protractor</u> to find out how large it is in <u>degrees</u>.

2) Then <u>divide</u> that number by <u>360</u> and <u>times</u> by <u>100</u>.

3) To find the <u>amount</u> a wedge of the pie is <u>worth</u>, work out your <u>percentage</u> then turn it into a <u>decimal</u>. Then times the <u>decimal</u> by the <u>total amount</u> of the pie.

Pie Chart of Transport Type

Q: Out of 100 people, how many used a pogostick?
A: 126 – 90 = 36°, so (36 ÷ 360) × 100 = 10%, so 0.1 × 100 = <u>10 people</u>.

2 Completing Pie Charts

1) To <u>draw</u> on a <u>new wedge</u> that you know the <u>%</u> for, turn the % into a <u>decimal</u> and <u>times</u> it by <u>360</u>. Then draw a wedge of that many <u>degrees</u>.

Q: Out of 100 people, 25% used a bicycle. Add this to the pie chart.
A: 25 ÷ 100 = 0.25, 0.25 × 360 = <u>90°</u>.

2) To add a <u>new wedge</u> that you know the <u>amount</u> for, <u>divide</u> your amount by the <u>total amount</u> of the pie and <u>times</u> the answer by <u>360</u>. Then <u>draw</u> on a wedge of that many <u>degrees</u>.

Q: Out of 100 people, 55 used a car, add this to the pie chart.
A: 55 ÷ 100 = 0.55, 0.55 × 360 = <u>198°</u> (198° + 126° = <u>324°</u>).

Sorry darling, we've got no relationship — look at our scatter graph...

Hmm, who'd have thought <u>pie</u> could be so complicated. Don't panic though, a bit of <u>practice</u> and you'll be fine. And don't worry, you're over half way through this section now. Congratulations — I'm so proud of you, sniff.

Maps

A couple of jazzy maps on this page for you, both with complicated names — topological and proportional symbol. And a bit on isolines too. You need to learn what they're for and how they work.

Topological Maps are Simplified Maps

1) Some maps are hard to read because they show too much detail.

2) Topological maps get around this by just showing the most important features like roads and rail lines. They don't have correct distances or directions either, which makes them easier to read.

3) They're often used to show transport networks, e.g. the London tube map.

4) If you have to read a topological map — dots are usually places and lines usually show routes between places. If two lines cross at a dot then it's usually a place where you can switch routes.

5) As always, don't forget to check out the key.

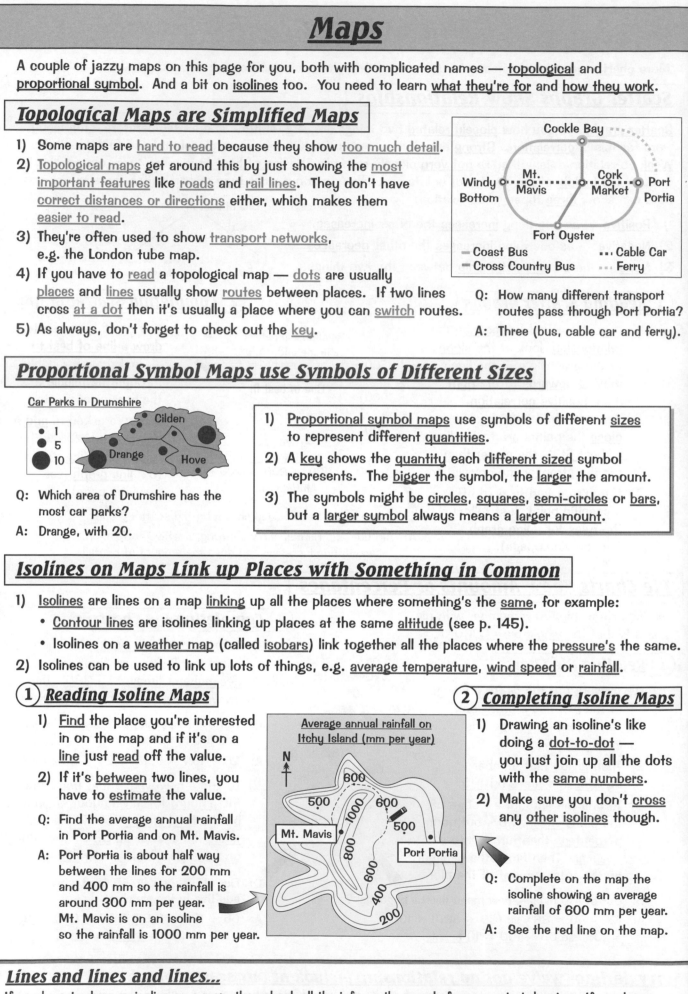

Cockle Bay

Windy Bottom   Mt. Mavis   Cork Market   Port Portia

Fort Oyster

— Coast Bus   ... Cable Car
— Cross Country Bus ... Ferry

Q: How many different transport routes pass through Port Portia?

A: Three (bus, cable car and ferry).

Proportional Symbol Maps use Symbols of Different Sizes

Car Parks in Drumshire

• 1
• 5
● 10

Cilden
Drange   Hove

Q: Which area of Drumshire has the most car parks?

A: Drange, with 20.

1) Proportional symbol maps use symbols of different sizes to represent different quantities.

2) A key shows the quantity each different sized symbol represents. The bigger the symbol, the larger the amount.

3) The symbols might be circles, squares, semi-circles or bars, but a larger symbol always means a larger amount.

Isolines on Maps Link up Places with Something in Common

1) Isolines are lines on a map linking up all the places where something's the same, for example:
   • Contour lines are isolines linking up places at the same altitude (see p. 145).
   • Isolines on a weather map (called isobars) link together all the places where the pressure's the same.

2) Isolines can be used to link up lots of things, e.g. average temperature, wind speed or rainfall.

① Reading Isoline Maps

1) Find the place you're interested in on the map and if it's on a line just read off the value.

2) If it's between two lines, you have to estimate the value.

Q: Find the average annual rainfall in Port Portia and on Mt. Mavis.

A: Port Portia is about half way between the lines for 200 mm and 400 mm so the rainfall is around 300 mm per year. Mt. Mavis is on an isoline so the rainfall is 1000 mm per year.

Average annual rainfall on Itchy Island (mm per year)

N

600
500   600
1000
Mt. Mavis   500
800
Port Portia
600
400
200

② Completing Isoline Maps

1) Drawing an isoline's like doing a dot-to-dot — you just join up all the dots with the same numbers.

2) Make sure you don't cross any other isolines though.

Q: Complete on the map the isoline showing an average rainfall of 600 mm per year.

A: See the red line on the map.

Lines and lines and lines...

If you have to draw an isoline on a map, then check all the info on the map before you start drawing. If you know where the line's got to go you won't muck it up. Make sure you do it in pencil too, so you can rub out any mistakes.

Maps

Three more maps, with three more ludicrous names. Well the last two aren't that bad, but this first one — <u>choropleth</u>, sounds like a treatment at the dentist.

Choropleth Maps show How Something Varies Between Different Areas

1) <u>Choropleth maps</u> show how something varies between different areas using <u>colours</u> or <u>patterns</u>.

2) The maps in exams often use <u>cross-hatched lines</u> and <u>dot patterns</u>.

3) If you're asked to talk about all the parts of the map with a certain <u>value</u> or <u>characteristic</u>, look at the map carefully and put a <u>big tick</u> on all the parts with the <u>pattern</u> that <u>matches</u> what you're looking for. This makes them all <u>stand out</u>.

4) When you're asked to <u>complete</u> part of a map, first use the <u>key</u> to work out what type of <u>pattern</u> you need. Then <u>carefully</u> draw on the pattern, e.g. using a <u>ruler</u>.

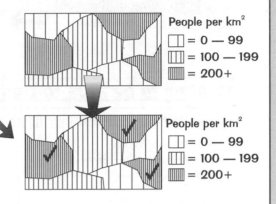

People per km²
☐ = 0 — 99
▥ = 100 — 199
▦ = 200+

People per km²
☐ = 0 — 99
▥ = 100 — 199
▦ = 200+

Flow Lines show Movement

1) <u>Flow line maps</u> have <u>arrows</u> on, showing how things <u>move</u> (or are moved) from one place to another.

2) They can also be <u>proportional symbol maps</u> — the <u>width</u> of the arrows show the <u>quantity</u> of things that are <u>moving</u>.

Q: From which <u>area</u> do the <u>greatest</u> number of people entering the UK come from?

A: <u>USA</u>, as this arrow is the largest.

Q: The number of people entering the UK from the <u>Middle East</u> is <u>roughly half</u> the number of people entering from the <u>USA</u>. Draw an <u>arrow</u> on the map to <u>show</u> this.

A: Make sure your arrow is going in the <u>right direction</u> and its <u>size</u> is appropriate (e.g. <u>half the width</u> of the USA arrow).

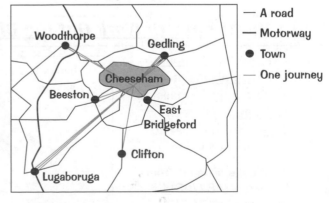

Some of the flows of people to the UK

USA

Middle East

Rest of the Americas

Immigration

Desire Lines show Journeys

1) <u>Desire line maps</u> are a type of flow line as they show <u>movement</u> too.

2) They're <u>straight lines</u> that show <u>journeys</u> <u>between</u> two <u>locations</u>, but they <u>don't follow</u> <u>roads</u> or <u>railway lines</u>.

3) <u>One line</u> represents <u>one journey</u>.

4) They're used to show <u>how far</u> all the people have <u>travelled</u> to get to a <u>place</u>, e.g. a shop or a town centre, and <u>where</u> they've <u>come from</u>.

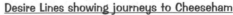

Desire Lines showing journeys to Cheeseham

— A road
— Motorway
● Town
— One journey

Woodthorpe
Gedling
Cheeseham
Beeston
East Bridgeford
Clifton
Lugaboruga

Desire lines — I'm sure my palm reader mentioned those...

...unfortunately I'm not as good at seeing the future as she is* so I can't predict if any of these <u>maps</u> are going to come up in your <u>exam</u>. They could do though, so make sure you know what they are and how to read them.

*If you're wondering, I'm going to meet a tall, dark, handsome stranger very soon....

Exam Skills

Ordnance Survey Maps

Next up, the dreaded <u>Ordnance Survey®</u> <u>maps</u>. Don't worry, they're easy once you know how to use 'em.

Learn These Common Symbols

Ordnance survey (OS®) maps use lots of <u>symbols</u>. It's a good idea to learn some of the most <u>common ones</u> — like these:

▬▬ Motorway	─·─ County boundary	PO	Post Office®
▬▬ Main (A) road	═══ National Park boundary	PH	Pub
═══ Secondary (B) road	---- Footpath	+	Place of worship
⋈ Bridge	▨ Building	♦	Place of worship, with a tower
─── Railway	●─ Bus station	♦	Church with a spire, minaret or dome

You have to be able to Understand Grid References

You need to be able to use <u>four figure</u> and <u>six figure</u> <u>grid references</u> for your exam.

Q: Give the four figure and six figure grid reference for the Post Office®.

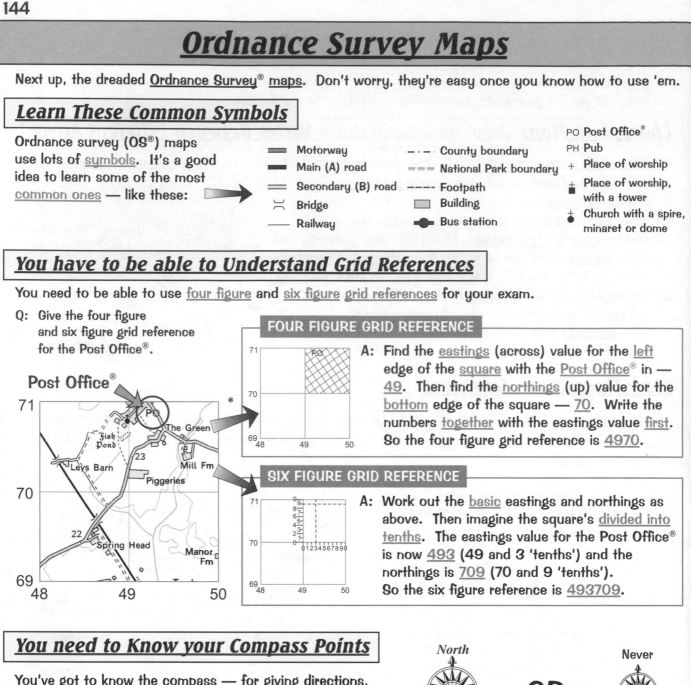

Post Office®

FOUR FIGURE GRID REFERENCE

A: Find the <u>eastings</u> (across) value for the <u>left</u> edge of the <u>square</u> with the <u>Post Office®</u> in — <u>49</u>. Then find the <u>northings</u> (up) value for the <u>bottom</u> edge of the square — <u>70</u>. Write the numbers <u>together</u> with the eastings value <u>first</u>. So the four figure grid reference is <u>4970</u>.

SIX FIGURE GRID REFERENCE

A: Work out the <u>basic</u> eastings and northings as above. Then imagine the square's <u>divided into</u> <u>tenths</u>. The eastings value for the Post Office® is now <u>493</u> (49 and 3 'tenths') and the northings is <u>709</u> (70 and 9 'tenths'). So the six figure reference is <u>493709</u>.

You need to Know your Compass Points

You've got to know the compass — for giving <u>directions</u>, saying <u>which way</u> a <u>river's flowing</u>, or knowing what they mean if they say 'look at the river in the <u>NW</u> of the map' in the exam. Read it <u>out loud</u> to yourself, going <u>clockwise</u>.

North
West — East
South

OR

Never
Wheat — Eat
Soggy

You Might have to Work Out the Distance Between Two Places

To work out the <u>distance</u> between <u>two places</u> on a <u>map</u>, use a <u>ruler</u> to measure the <u>distance</u> in <u>cm</u> then <u>compare</u> it to the scale to find the distance in <u>km</u>.

Q: What's the distance from the bridge (482703) to the church (490708)?

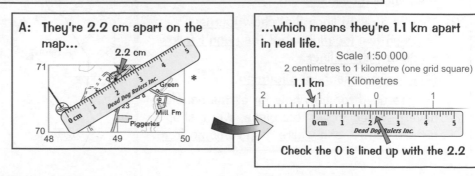

A: They're 2.2 cm apart on the map...

2.2 cm

Dead Dog Rulers Inc.

...which means they're 1.1 km apart in real life.

Scale 1:50 000
2 centimetres to 1 kilometre (one grid square)

1.1 km   Kilometres

Dead Dog Rulers Inc.

Check the 0 is lined up with the 2.2

Keeping ramblers happy since 1791...

I told you <u>OS maps</u> aren't as bad as you thought. If a dodgy looking rambler who's been walking in the rain for five hours with only a cup of tea to keep him going can read them, then so can you. Get ready for some more <u>map</u> fun...

Exam Skills

Ordnance Survey Maps

Almost done with <u>exam skills</u> now. Just this final page looking at <u>contour lines</u> and <u>sketching</u> from Ordnance Survey® maps or photographs to deal with then you're free, free I tell you...

The Relief of an Area is Shown by Contours and Spot Heights

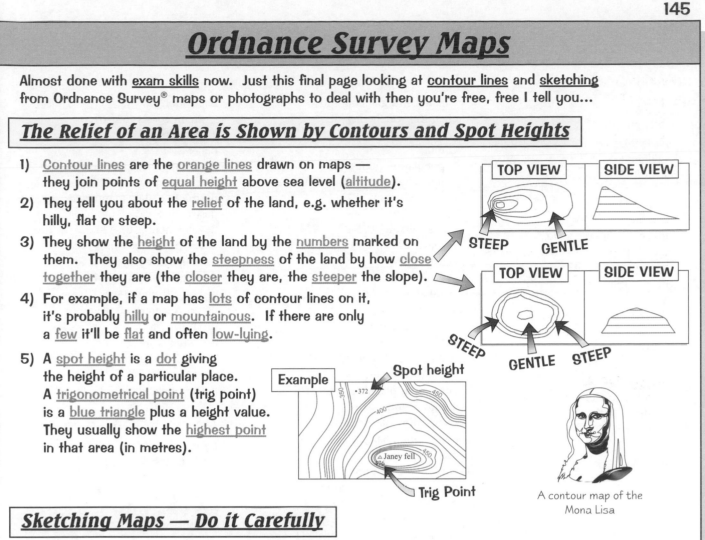

1) <u>Contour lines</u> are the <u>orange lines</u> drawn on maps — they join points of <u>equal height</u> above sea level (<u>altitude</u>).

2) They tell you about the <u>relief</u> of the land, e.g. whether it's hilly, flat or steep.

3) They show the <u>height</u> of the land by the <u>numbers</u> marked on them. They also show the <u>steepness</u> of the land by how <u>close together</u> they are (the <u>closer</u> they are, the <u>steeper</u> the slope).

4) For example, if a map has <u>lots</u> of contour lines on it, it's probably <u>hilly</u> or <u>mountainous</u>. If there are only a <u>few</u> it'll be <u>flat</u> and often <u>low-lying</u>.

5) A <u>spot height</u> is a <u>dot</u> giving the height of a particular place. A <u>trigonometrical point</u> (trig point) is a <u>blue triangle</u> plus a height value. They usually show the <u>highest point</u> in that area (in metres).

TOP VIEW SIDE VIEW
STEEP GENTLE

TOP VIEW SIDE VIEW
STEEP GENTLE STEEP

Example
Spot height •372
Trig Point △ Janey fell

A contour map of the Mona Lisa

Sketching Maps — Do it Carefully

1) In the <u>exam</u>, they could give you a <u>map</u> or <u>photograph</u> and tell you to <u>sketch</u> part of it.

2) Make sure you figure out <u>what bit</u> they want you to sketch out, and <u>double check</u> you've <u>got it right</u>. It might be only <u>part</u> of a lake or a wood, or only <u>one</u> of the roads.

3) If you're <u>sketching</u> an <u>OS</u>® <u>map</u>, it's a good idea to <u>copy</u> the <u>grid</u> from the map onto your sketch paper — this helps you to copy the map <u>accurately</u>.

4) Draw your sketch <u>in pencil</u> so you can <u>rub it out</u> if it's <u>wrong</u>.

5) Look at how much <u>time</u> you have and <u>how many marks</u> it's worth to decide how much <u>detail</u> to add.

Q: Draw a labelled sketch of the OS map shown below.

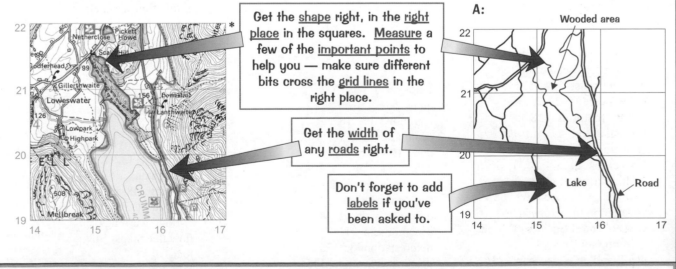

Get the <u>shape</u> right, in the <u>right place</u> in the squares. <u>Measure</u> a few of the <u>important points</u> to help you — make sure different bits cross the <u>grid lines</u> in the right place.

Get the <u>width</u> of any <u>roads</u> right.

Don't forget to add <u>labels</u> if you've been asked to.

A:
Wooded area
Lake Road

What a relief that's over...

When you're <u>sketching</u> a copy of a map or photo see if you can lay the paper over it — then you can <u>trace</u> it (sneaky). And that my friends is the end of the <u>exam skills</u> section, and the <u>end of the book</u>. Now go treat yourself to an exam.

* Maps: Reproduced from Ordnance Survey digital map data © Crown copyright 2001

Exam Skills

Index

Index

Index